SERIES

ENTOMOLOGICA

EDITORS

E. SCHIMITSCHEK & K. A. SPENCER

VOLUME 16

Dr W. Junk bv Publishers The Hague–Boston–London 1979

© Dr W. Junk bv – Publishers – The Hague 1979
Softcover reprint of the hardcover 1st edition 1979

Cover design Max Velthuijs
ISBN-13: 978-94-009-9602-1   e-ISBN-13: 978-94-009-9600-7
DOI: 10.1007/978-94-009-9600-7

SKELETAL MUSCULATURE IN LARVAL PHASES OF THE BEETLE *EPICAUTA SEGMENTA* (COLEOPTERA, MELOIDAE)

by

ANGEL BERRÍOS-ORTIZ and RICHARD B. SELANDER

Dr W. Junk bv Publishers The Hague–Boston–London 1979

Library of Congress Cataloging in Publication Data   CIP

Berríos-Ortiz, Angel.
    Skeletal musculature in larval phases of the beetle Epicauta segmenta (Coleoptera,
Meloidae).

    (Series entomologica; v. 16)
    Bibliography: p.
    Includes index.
    1. Epicauta segmenta—Anatomy.   2. Larvae—Insects.
3. Muscles.   4. Insects—Anatomy.
I. Selander, Richard Brent, 1927–   joint author.   II.   Title.
QL596.M38B47   595.7′65   79–12114

Angel Berríos-Ortiz: Departamento de Biología, Universidad de Puerto Rico en
Mayagüez, Mayagüez, Puerto Rico 00708

Richard B. Selander: Department of Genetics and Development, University of
Illinois, Urbana, Illinois 61801

CONTENTS

INTRODUCTION

This study, the first in an intended series of anatomical investigations of the blister beetles, was undertaken primarily for the purpose of determining the changes that occur in the skeletal musculature during postembryonic larval development. The species studied, *Epicauta segmenta* (Say), like others belonging to the coleopterous family Meloidae, is characterized by hypermetamorphosis (SELANDER and WEDDLE, 1969). The egg develops into an active, well sclerotized larva that searches for grasshopper eggs, which, as in the case of all species of *Epicauta*, serve as the sole larval food. This *triungulin phase* of the larval stage, as it is called, is followed by the *first grub phase* (Fig. 1), during which the larva continues to feed and undergoes several molts. After feeding is completed the larva commonly enters a sessile, diapausing *coarctate phase* in which the integument is heavily sclerotized and the appendages, including the legs, are vestigial (Fig. 2). Following this the larva attains an active *second grub phase* (Fig. 3) closely resembling the first grub phase in external anatomy. Normally, the second grub phase leads directly to pupation and the formation of the adult. An alternate, abbreviated developmental pattern, involving pupation immediately after the first grub phase, is also recorded in this and many other species of *Epicauta*.

Although it has long been known that larval Meloidae undergo marked modifications of external anatomy, behavior, and ecology in progressing through the various larval phases (hence the name hypermetamorphosis), little or no attention has been paid to differences among phases with respect to internal anatomical characters. Consequently, the present work, which treats the gross comparative anatomy of the first grub, coarctate, and second grub phases of *E. segmenta*, should be of considerable interest and value to students of the biological mechanisms and evolutionary significance of hypermetamorphosis. Included within the scope of the study are all muscles attached to the exoskeleton except for intrinsic muscles of the appendages.

Literature Review

The anatomy of larval insects has been studied in the past largely as an adjunct of taxonomy (*e.g.*, PETERSON, 1951–1962). In particular, because larval musculature is not very conveniently employed for taxonomic purposes, it has received much less attention than external characters of the larval stage. CRAMPTON (1918, 1921), for example, does not refer to musculature at all in his general study of the head and thorax of immature insects. DAS (1937) makes an attempt to survey the musculature of larval mouthparts and suggests the possiblility of

1

using musculature as a basis for the classification of insects. SNODGRASS (1931) deals with larval musculature only in very broad terms.

Larvae of Coleoptera have been studied anatomically by many of our most competent entomologists. For example, BLAIR (1933), BÖVING (1930), CARPENTER (1912), HAYES (1927, 1928, 1929), HURPIN (1953), MARCUZZI (1960), PERRIS (1877), ROBERTS (1930), VERHOEFF (1921), and WHITEHEAD (1932) present extensive descriptions of the external anatomy of a wide variety of coleopterous larvae. But in general, the anatomical aspect of these and similar works is secondary to the taxonomic. More pertinent to the present study are those of BERLESE (1909), BÖVING (1914), BREED (1903), CRAIGHEAD (1916), CROME (1957), JÖSTING (1942), MURRAY and TIEGS (1935), PATERSON (1930), PADHAN (1949), SPEYER (1922), and TIEGS and MURRAY (1938), which treat the musculature of larval Coleoptera in greater or lesser detail. JÖSTING's work is unique in relating specifically to muscular innervation (in *Tenebrio*). Mention should also be made of studies of the musculature of larval mouthparts by ANDERSON (1936) and DORSEY (1943).

Although larval and adult external features are, of course, routinely described in the taxonomic literature of the Meloidae, there are few works devoted to anatomy *per se*. BLOOD (1935) describes the adult anatomy of *Pyrota mylabrina* (Chevrolat). BRETHÉS (1917) treats some aspects of the adult anatomy of *Epicauta adspersa* (Klug). EVERLY (1936) and JUDD (1947) consider the anatomy of the alimentary tracts of the adults of two species of *Epicauta*. SAXENA (1953, 1955) presents a thorough study of the exoskeleton and musculature of a species of *Mylabris*, but again only the adult stage is considered. Lately, GERBER, CHURCH, and REMPEL (1971, 1972) and CHURCH and REMPEL (1971) have done elegant work on the embryology, histology, and anatomy of the reproductive systems (male and female) in *Lytta nuttalli* (Say).

Descriptions of the external anatomy of larval Meloidae of more than ordinary thoroughness and interest appear in the works of BEAUREGARD (1890), CROS (1917, 1918, 1919, 1924a, 1924b, 1926, 1928), HACHFIELD (1928), HORSFALL (1943), MacSWAIN (1956), MAYET (1875), MILLIKEN (1921), NAGATOMI and IWATA (1958), NEWPORT (1853), PARDO ALCAIDE (1953), RILEY (1877), SELANDER and MATHIEU (1964), STEINKE (1919), and YAKHONTOV (1931). The larval musculature system of the Meloidae, however, remains virtually unknown.

Acknowledgments

We are grateful to CHERYL ADAMS, JOHN BOUSEMAN, JOHN BOYD, and NANCY AGAFITEI for their encouragement and assistance. The project was supported by grants (GB-5547 and GB-35516) from the National Science Foundation (R. B. SELANDER, Principal Investigator).

MATERIAL AND METHODS

Epicauta segmenta is one of eight species of the Albida Group of the subgenus *Macrobasis*. The geographic range of the species extends from northern Mexico through much of the southwestern United States to a northern limit in South Dakota. All material used in the present study was reared in the laboratory from a stock of adult beetles collected at Fort Davis, Texas. The technique utilized in the rearings is described by SELANDER and WEDDLE (1969). Material selected for dissection was killed and fixed in alcoholic Bouin's solution, in which it remained for a day before transfer to a 70% solution of ethyl alcohol in water for preservation.

Most of the larvae utilized were reared specifically for anatomical study, although some of the coarctate larvae were obtained from completed experimental rearings. The description of musculature in the first grub phase is based on larvae in the fifth instar, which is usually the last instar in that phase (SELANDER and WEDDLE, 1969). Larvae in the first and second grub phases were killed and fixed several days after ecdysis. Coarctate larvae had been in that phase for several days to more than a year at the time they were killed and fixed. Following the system of SELANDER and MATHIEU (1964), the symbols FG, C, and SG will be used to designate the first grub, coarctate, and second grub phases, respectively.

Dissections were made from the dorsal, sternal, and lateral regions under a dissecting miscroscope. A Petri dish lined with paraffin wax to which ground carbon had been added was used to hold larvae during dissection. Specimens were totally immersed in distilled water to which a few drops of a saturated aqueous solution of methylene blue were added to give contrast to the tissues. In this medium the muscles take on a satin sheen, the nerves appear silky, and the fat bodies are nicely differentiated by their dull appearance.

Drawings were made directly from dissections using an ocular grid. In order to avoid excessive complication and possible confusion in the drawings, only a single set or layer of muscles is shown in most of them. In any series of drawings representing a general view (*e.g.*, frontal, lateral, sagittal) of a section of the head, thorax, or abdomen, consecutively numbered figures show muscles lying progressively farther from the viewer. The sets of muscles shown in individual figures are not necessarily functional sets, of course.

Because the musculature is completely symmetrical bilaterally, it is both convenient and sufficient, in describing the musculature, to refer to only a single side. Thus, for example, specification of *six* cibarial muscles in the head implies two bilaterally symmetrical sets of six muscles. Muscles are designated according to the system proposed by CHADWICK (1957). Numerical subscripts designate individual thoracic and abdominal segments, the Arabic numerals 1 through 3

3

are used for the thoracic segments and the Roman numerals I through IX for the abdominal segments. Like the body segments, intersegmental membranes are numbered consecutively from anterior to posterior, but in this case no distinction is made between the thorax and abdomen. The sequence of Arabic numerals 0 through 12 is used for the membranes, membrane 0 being that between the head and thorax and 12 that between abdominal segments VIII and IX.

The convention adopted in lettering a set of similar muscles is as follows: Lettering is alphabetical from the dorsal midline of the body for dorsal muscles, from the ventral midline for ventral muscles, and from anterior to posterior for lateral muscles. When the muscles of a set differ in distance from the median axis of the body, the lettering begins with the innermost layer. Although the drawings portray the musculature quite realistically, some schematization was unavoidable. Only labelled muscles are cited in references to figures in the text.

Descriptions of muscles follow the general pattern of specifying point of origin first and insertion second. Thus, for example, 'parietal wall of the cranium, laterad of the adrontal area, to the posterior margin of the base of the scape' is to be read as 'originating on the parietal wall of the cranium, laterad of the adfrontal area, and inserting on the posterior margin of the base of the scape'. Unless specifically noted otherwise, descriptions refer to all three larval phases treated (FG, C, and SG). As will be seen, the pattern of musculature is virtually identical in these phases.

GLOSSARY OF ABBREVIATIONS

The following abbreviations are used consistently in the text and figures:

cd: cardo
clp: clypeus
cx: coxa
dmb: dorsal intersegmental membrane
ephy: epipharynx
fl: flagellum
fr: frons
ga: galea
ge: gena
hphy: hypopharynx
lb: labium
lc: lacinia
lm: labrum
lmb: lateral intersegmental membrane
lvmb: lateroventral intersegmental membrane
md: mandible
mx: maxilla
oc: occiput
pdc: pedicel
phy: pharynx
pl: pleuron
plp: palpus
pmt: postmentum
por: postoccipital ridge
prmt: prementum
prtl: parietal
sc: scape
sp: spiracle
St: stipes
st: sternum
t: tergum
tnt: tentorium
tr: trochanter
vmb: ventral intersegmental membrane

MUSCLES OF THE HEAD

The musculature of the head is composed of muscles associated with the antennae, pharynx, cibarium, and mouthparts. The total number of muscles in each side of the head is 21 in the FG and SG phases and 20 in the C phase. An inventory of the musculature is given in table 1.

Antennal. Two muscles are associated with each antenna (Figs. 5, 14–15, 25, 27, 32, 34, 41, 44, 52, 55, 65).

prtl–scp$_a$: Parietal wall of the cranium, laterad of the adfrontal area, to the anterior margin of the base of the scape.

prtl–scp$_b$: Parietal wall of the cranium, laterad of the adfrontal area, to the posterior margin of the base of the scape.

Labral. One muscle (Figs. 5, 14, 20–21, 27, 32, 37–38, 44, 53, 59–60).

fr–lm: Posterior part of the frons laterally to the posterior margin of the base of the labrum.

Pharyngeal–cibarial. Six muscles insert on the pharynx or walls of the cibarium (Figs. 5–6, 8, 12–14, 18–24, 27–28, 30, 32, 36–40, 44–45, 50–53, 57–62).

clp–ephy$_a$ Clypeal region laterally to the epipharyngeal wall of the cibarium.

clp–ephy$_b$ Clypeal region, posteriad of the origin of *clp–ephy$_a$*, to the middle of the epipharyngeal wall of the cibarium. Formed by two bundles.

fr–phy$_a$: Frons to the dorsal wall of the pharynx posteriad of the frontal ganglion connective and mesad of *fr–hphy*.

fr–phy$_b$: Frons to the posterior margin of the dorsal wall of the pharynx, posteriad of *fr–phy$_a$* and *fr–hphy*. Formed by three bundles.

fr–hphy: A branched muscle running from the frons to the dorsal end of the hypopharyngeal bar.

oc–phy: Lateral region of the occiput to the lateral wall of the pharynx, inserting posteriad of *fr–phy$_b$*.

Mandibular. Two large, multibranched muscles are associated with each mandible (Figs. 6–7, 12–13, 15–16, 24–26, 27, 31, 34–35, 40–41, 45–46, 51, 55–56, 63–64). Each muscle inserts on an apodeme.

prtl–md$_a$: Lateral wall of the cranium to the lateral margin of the mandible.

prtl–md$_b$: Lateral and dorsal walls of the cranium mesad of prtl–md$_a$ to the mesal angle of the mandibular base.

Maxillary. Seven muscles are associated with each maxilla (Figs. 8–9, 11–12, 17–18, 22–23, 25, 28, 30, 35–36, 39–41, 47–48, 50–51, 57–58, 62–64).

tnt–cd: Base of the posterior tentorial arm to the mesoventral margin of the base of the cardo.

tnt–St$_a$: Tentorium to the posterior margin of the base of the stipes.

tnt–St$_b$: Tentorium to the lateroventral margin of the stipes.

tnt–St$_c$: Similar to, and inserting slightly laterad of, *tnt–St$_b$*.

tnt–St$_d$: Similar to, and inserting slightly laterad of, *tnt–St$_c$*.

oc–St: A long, double muscle from the lateroventral region of the occiput to the middorsal region of the stipital base.

St–plp: Mesal edge of the stipes to the ventral margin of the basal segment of the palpus. Absent in the coarctate phase.

Labial. Three muscles (Figs. 9, 11–12, 18–19, 21, 28, 30, 37–38, 47–48, 50–51, 59–60).

tnt–prmt$_a$: Base of the posterior tentorial arm to the dorsal margin of the prementum.

tnt–prmt$_b$: Base of the posterior arm of the tentorium to the ventral margin of the prementum, mesad of *tnt–prmt$_a$*.

pmt–prmt: Posterior region of the postmentum to the base of the prementum.

Table 1. Inventory of muscles of the head in the larval phases studied (+ means present, − absent).

Muscles (by group)	Phases		
	FG	C	SG
Antennal			
prtl-scp$_a$	+	+	+
prtl–scp$_b$	+	+	+
Labral			
fr–lm	+	+	+
Pharyngeal–cibarial			
clp–ephy$_a$	+	+	+
clp–ephy$_b$	+	+	+
fr–phy$_a$	+	+	+
fr–phy$_b$	+	+	+
fr–hphy	+	+	+
oc–phy	+	+	+
Mandibular			
prtl–md$_a$	+	+	+
prtl–md$_b$	+	+	+
Maxillary			
tnt–cd	+	+	+
tnt–St$_a$	+	+	+
tnt–St$_b$	+	+	+
tnt–St$_c$	+	+	+
tnt–St$_d$	+	+	+
oc–St	+	+	+
St-plp	+	−	+
Labial			
tnt–prmt$_a$	+	+	+
tnt–prmt$_b$	+	+	+
pmt–prmt	+	+	+

MUSCLES OF THE THORAX

The prothorax, mesothorax, and metathorax are treated separately in the following sections. There are, in each complete lateral set, 48 prothoracic, 48 mesothoracic, and 47 metathoracic muscles (table 2).

Prothorax

Dorsal longitudinal. Eleven muscles constitute this series (Figs. 68–69, 71–74, 77–78, 86–87, 90–93, 97–98, 105–106, 109–111, 113–116, 120–123, 132–133). One muscle (t_1-por_a) has no homolog in the mesothorax or metathorax. In addition, the prothorax lacks the homologs of $1dmb-2dmb_g$ and $2dmb-3dmb_g$ of the mesothorax and metathorax, respectively.

$1dmb-por_a$: First dorsal intersegmental membrane (between the prothorax and mesothorax) to the dorsomedian region of the postoccipital ridge.

$1dmb-por_b$: Similar to, and inserting laterad of, $1dmb-por_a$.

$1dmb-por_c$: Similar to $1dmb-por_a$; laterad of $1dmb-por_b$.

$1dmb-por_d$: Similar to $1dmb-por_a$; inserting laterad of $1dmb-por_c$.

$1dmb-por_e$: Similar to $1dmb-por_a$; inserting laterad of $1dmb-por_d$.

$1dmb-por_f$: A transverse muscle from the first dorsal intersegmental membrane to the dorsolateral region of the postoccipital ridge.

t_1-por_a: A transverse, branched muscle from the protergum to the dorsolateral region of the postoccipital ridge.

$1dmb-t_{1a}$: First dorsal intersegmental membrane to the lateral portion of the protergum.

$1dmb-t_{1b}$: Similar to $1dmb-t_{1a}$.

$1dmb-t_{1c}$: Similar to $1dmb-t_{1a}$; inserting ventrad of $1dmb-t_{1b}$.

$1dmb-t_{1d}$: Similar to $1dmb-t_{1a}$; inserting ventrad of $1dmb-t_{1c}$.

Ventral longitudinal. Six muscles form this series (Figs. 69–72, 79–80, 84–85, 90–92, 99, 103–104, 111–114, 124–125, 129–130).

$1vmb-por_a$: First ventral intersegmental membrane to the ventral region of the postoccipital ridge near the anterior end of the prosternal plate.

$1vmb-por_b$: Similar to $lvmb-por_a$; inserting dorsad of $1vmb-por_a$.

$1vmb-por_c$: Similar to $1vmb-por_a$; inserting laterad of $1vmb-por_a$.

$1vmb-por_d$: Similar to $1vmb-por_a$; inserting dorsad of $1vmb-por_c$.

$1lvmb-por_a$: Lateroventral region of the first ventral intersegmental membrane to the ventral region of the postoccipital ridge.

$1lvmb-por_b$. A thin paraventral muscle, from the lateroventral region of the first ventral intersegmental membrane to the ventral region of the postocciptal ridge.

9

Oblique intersegmental. There are five muscles in this series (Figs. 69–74, 79, 81, 83, 85–86, 90–94, 98, 100, 103–104, 110–116, 123, 125, 128–129).

st_1–por_a: A thin muscle, from the anterior portion of the prosternal plate to the lateral region of the postoccipital ridge.

st_1–por_b: A heavy muscle, from the posterior part of the prosternal plate to the lateral region of the postoccipital ridge.

st_1–por_c: A heavy muscle, from the prosternum to the lateral region of the postoccipital ridge laterad of st_1–por_a.

$1dmb$–por_g: Lateral region of the first dorsal intersegmental membrane to the ventral region of the postoccipital ridge.

$1dmb$–por_h: Similar to $1dmb$–por_g.

Dorsoventral. This series consists of 15 muscles, most of which are very thin (Figs. 67–71, 73–75, 78–80, 83–84, 86–87, 89–94, 98–100, 104, 108–110, 112–113, 116–117, 121–124, 131).

t_1–st_1: Lateral region of the protergum to the anterolateral region of the prosternum.

t_1–$1vmb$: Posterolateral region of the protergum to the first ventral intersegmental membrane.

t_1–tr_{1a}: Lateral region of the protergum to the anterodorsal margin of the protrochanter.

t_1–tr_{1b}: Similar to, and inserting posteriad of, t_1–tr_{1a}.

t_1–tr_{1c}: Lateral region of the protergum to the dorsal margin of the protrochanter posteriad of t_1–tr_{1b}.

t_1–tr_{1d}: Lateral region of the protergum to the posterodorsal margin of the protrochanter posteriad of t_1–tr_{1c}.

t_1–tr_{1e}: Lateral region of the protergum to the laterodorsal margin of the protrochanter.

t_1–tr_{1f}: Similar to, and inserting posteriad of, t_1–tr_{1e}.

t_1–tr_{1g}: Similar to t_1–tr_{1e}; inserting posteriad of t_1–tr_{1f}.

t_1–tr_{1h}: Similar to t_1–tr_{1e}; inserting posteriad of t_1–tr_{1g}.

t_1–cx_{1a}: Protergum to the anterodorsal region of the procoxa.

t_1–cx_{1b}: Similar to t_1–cx_{1a}.

t_1–cx_{1c}: Similar to, and inserting laterad of, t_1–cx_{1a}.

t_1–cx_{1d}: Similar to t_1–cx_{1a}.

t_1–cx_{1e}: Similar to t_1–cx_{1a}.

Sternopleural. There are no sternopleural muscles in the prothorax.

Tergopleural. There are six muscles in this series (Figs. 67, 69, 72, 74–75, 77, 79–80, 84–86, 89, 91, 94, 97–98, 103, 108, 111, 115, 117–118, 122, 129–130, 132).

t_1–por_b: A branched muscle from the lateral margin of the protergum to the ventrolateral portion of the postoccipital ridge.

t_1–pl_{1a}: Protergum to the posterodorsal margin of the pleural region.

t_1–pl_{1b}: Protergum to the posterodorsal margin of the propleuron.

t_1–pl_{1c}: Similar to t_1–pl_{1a}.

t_1–pl_{1d}: Similar to t_1–pl_{1a}.

t_1–pl_{1e}: Similar to t_1–pl_{1a}.

Pleurocoxal. There are no pleurocoxal muscles in the prothorax.

Sternocoxal. There are no sternocoxal muscles in the prothorax.

Sternotrochanteral. Three muscles constitute this series (Figs. 67–68, 73–74, 81, 83, 89, 94, 100, 103, 109, 117, 126, 128).

$1vmb$–tr_{1a}: First ventral intersegmental membrane to the posteroventral edge of the protrochanter.

$1vmb$–tr_{1b}: First ventral intersegmental membrane to the anterodorsal edge of the protrochanter.

$1vmb$–tr_{1c}: Similar to $lvmb$–tr_{1b}.

Spiracular. There are two muscles associated with each prothoracic spiracle. These muscles are not illustrated.

sp_1–pl_1: From the first (prothoracic) spiracle to the posteropleural region of the prothorax.

sp_1–cx_1: First spiracle to the posterodorsal edge of the procoxa.

Mesothorax

Dorsal longitudinal. There are 11 dorsal longitudinal muscles in the mesothorax (Figs. 68–69, 71–72, 77–78, 86–87, 90–91, 97–98, 105–106, 109–111, 113–115, 120–123, 130–131).

$1dmb$–$2dmb_a$: Anterior margin of the first dorsal intersegmental membrane to the anterior edge of the second intersegmental membrane (between the mesathorax and metathorax).

$1dmb$–$2dmb_b$: Similar to $1dmb$–$2dmb_a$.

$1dmb$–$2dmb_c$: Similar to $1dmb$–$2dmb_a$.

$1dmb$–$2dmb_d$: A paratergal muscle, from anterior edge of the first dorsal intersegmental membrane to the anterior edge of the second dorsal interseg-mental membrane.

$1dmb$–$2dmb_e$: An oblique muscle, from the first dorsal intersegmental mem-brane to the median region of the second dorsal intersegmental membrane.

$1dmb$–$2dmb_f$: Similar to $1dmb$–$2dmb_e$.

$1dmb$–$2dmb_g$: Similar to $1dmb$–$2dmb_f$.

t_2–$2dmb_a$: A thin muscle extending from the anterior margin of the mesoter-gum to the median region of the second dorsal intersegmental membrane.

t_2–$2dmb_b$: An oblique muscle, from the mesolateral region of the mesoter-gum to the second dorsal intersegmental membrane.

t_2–$2dmb_c$: An oblique muscle, from the lateral region of the mesotergum to the second dorsal intersegmental membrane.

t_2–$2dmb_d$: Similar to t_2–$2dmb_c$.

Ventral longitudinal. There are eight muscles in this series (Figs. 69–72, 79–80, 83–84, 90–93, 99, 103–104, 111–114, 124–125, 129–130).

$1vmb$–$2vmb_a$: First ventral intersegmental membrane to the second ventral intersegmental membrane.

$1vmb$–$2vmb_b$: Similar to $1vmb$–$2vmb_a$.

$1vmb$–$2vmb_c$: Similar to $1vmb$–$2vmb_a$.

11

1vmb–2vmb$_d$: Similar to *1vmb–2vmb$_a$*.

1vmb–2vmb$_e$: Similar to *1vmb–2vmb$_a$*.

1vmb–2vmb$_f$: An oblique muscle, from the first ventral intersegmental membrane to the second ventral intersegmental membrane.

1vmb–2vmb$_g$: Similar to *1vmb–2vmb$_f$*.

1lvmb–2lvmb: A parasternal muscle extending from the first lateroventral intersegmental membrane to the second lateroventral intersegmental membrane.

Oblique intersegmental. A single muscle (Figs. 70–71, 80, 85, 91–92, 99, 104, 112, 124, 131).

t$_2$–1vmb: A thin muscle extending from the anterolateral region of the metatergum to the first ventral intersegmental membrane.

Dorsoventral. Fifteen muscles constitute this series (Figs. 67, 69–75, 78, 80–81, 83–86, 89–94, 98–100, 103–104, 106, 108–113, 115–118, 121–124, 126, 129, 131).

t$_2$–st$_2$: Anterior laterodorsal region of the mesotergum to the lateral portion of the mesosternum.

t$_2$–2vmb: Posterolateral region of the mesotergum to the second ventral intersegmental membrane.

t$_2$–tr$_{2a}$: Anterior dorsolateral region of the mesotergum to the anterodorsal margin of the mesotrochanter.

t$_2$–tr$_{2b}$: Similar to *t$_2$–tr$_{2a}$*.

t$_2$–tr$_{2c}$: Similar to *t$_2$–tr$_{2a}$*.

t$_2$–tr$_{2d}$: Lateral region of the mesotergum to the laterodorsal margin of the mesotrochanter.

t$_2$–tr$_{2e}$: Similar to *t$_2$–tr$_{2d}$*.

t$_2$–tr$_{2f}$: Lateromedian region of the mesotergum to the laterodorsal margin of the mesotrochanter.

t$_2$–tr$_{2g}$: Lateroposterior region of the mesotergum to the laterodorsal margin of the mesotrochanter.

t$_2$–tr$_{2h}$: A thin muscle, from the lateroposterior margin of the mesotergum to laterodorsal margin of the trochanter.

t$_2$–cx$_{2a}$: Anterolateral margin of the mesotergum to the anterolateral edge of the mesocoxa.

t$_2$–cx$_{2b}$: Anterolateral region of the mesotergum to the anterodorsal margin of the mesocoxa.

t$_2$–cx$_{2c}$: Similar to *t$_2$–cx$_{2b}$*.

t$_2$–cx$_{2d}$: Lateral region of mesotergum to the anterodorsal margin of the mesocoxa.

t$_2$–cx$_{2e}$: A thin muscle, from the lateral portion of the mesotergum to the anterodorsal margin of the mexocoxa.

Sternopleural. There is a single sternopleural muscle in the mesothorax (Figs. 70–72, 79, 84, 90, 92, 99, 104, 112–113, 124, 130).

1vmb–2lmb: An oblique muscle, from the first ventral intersegmental membrane to the second lateral intersegmental membrane.

Tergopleural. There are six tergopleural muscles in the mesothorax (Figs.

67–68, 71–72, 74, 78–80, 85–86, 89–94, 98–99, 103–104, 108–109, 112–117, 124, 129–131).

t_2–pl_2: A thin muscle extending from the anterolateral margin of the mesotergum to the anterior ventrolateral margin of the mesopleuron.

$1dmb$–pl_{2a}: A wide, oblique muscle, from the dorsolateral region of the first intersegmental membrane to the posterior margin of the mesopleuron.

$1dmb$–$2pl_{2b}$: A short, wide, oblique muscle, from the posterior margin of the dorsolateral portion of the first dorsal intersegmental membrane to the anterior portion of the mesopleuron.

t_2–$2lmb_a$: Posterolateral portion of the mesotergum to the second lateral intersegmental membrane.

t_2–$2lmb_b$: Similar to t_2–$2lmb_a$.

t_2–$2lmb_c$: Similar to t_2–$2lmb_a$.

Pleurocoxal. One muscle (Figs. 67, 74, 80, 84, 89, 94, 100, 1ʹ , ---, -- , 129).

pl_2–cx_2: Posterior margin of the second pleuron to the anterodorsal margin of the mesocoxa.

Sternocoxal. One muscle (Figs. 67, 73, 84, 89, 94, 100, 109, 116, 126, 128).

$2vmb$–cx_2: Second ventral intersegmental membrane to the anteroventral margin of the mesocoxa.

Sternotrochanteral. There are four muscles in this series in the mesothorax (Figs. 68, 73–74, 81, 83, 89, 94, 100, 103, 109, 111, 115–116, 126, 128).

$2vmb$–tr_{2a}: Second ventral intersegmental membrane to the posteroventral margin of the mesotrochanter.

$2vmb$–tr_{2b}: A thin muscle, from the second ventral intersegmental membrane to the ventral margin of the mesotrochanter.

$2vmb$–tr_{2c}: A thin muscle, from the second ventral intersegmental membrane to the anterodorsal margin of the mesotrochanter.

$2vmb$–tr_{2d}: Similar to $2vmb$–tr_{2c}.

Spiracular. There are no spiracles on the mesothorax.

Metathorax

Dorsal longitudinal. Eleven muscles are represented in this series in the metathorax (Figs. 68–69, 71–72, 77–78, 86–87, 90–93, 97–98, 105–10ᶠ 109–111, 113–115, 120–123, 132–133).

$2dmb$–$3dmb_a$: Second dorsal intersegmental membrane medianlʏ
third dorsal intersegmental membrane (between the metathorax and abav.ɯɯnal segment I).

$2dmb$–$3dmb_b$: Second dorsal intersegmental membrane to the third dorsal intersegmental membrane.

$2dmb$–$3dmb_c$: Similar to $2dmb$–$3dmb_b$.

$2dmb$–$3dmb_d$: A paratergal muscle, from the second dorsal intersegmental membrane to the third dorsal intersegmental membrane.

$2dmb$–$3dmb_e$: An oblique muscle, from the second intersegmental membrane to the median region of the third dorsal intersegmental membrane.

13

$2dmb$–$3dmb_f$: Similar to $2dmb$–$3dmb_e$.

$2dmb$–$3dmb_g$: An oblique muscle, from the second intersegmental membrane to the lateromedian region of the third dorsal intersegmental membrane.

t_3–$3dmb_a$: A thin, median muscle, from the metatergum to the third intersegmental membrane.

t_3–$3dmb_b$: An oblique muscle, from the lateromedian region of the metatergum to the third intersegmental membrane.

t_3–$3dmb_c$: An oblique muscle, from the lateral region of the metatergum to the lateral portion of the third dorsal intersegmental membrane.

t_3–$3dmb_d$: Similar to t_3–$3dmb_c$.

Ventral longitudinal. Nine muscles (Figs. 69–72, 79–80, 83–85, 90–93, 99, 103–104, 111–114, 124–125, 129–130).

$2vmb$–$3vmb_a$: Median region of the second ventral intersegmental membrane to the third ventral intersegmental membrane.

$2vmb$–$3vmb_b$: Second ventral intersegmental membrane to the third ventral intersegmental membrane.

$2vmb$–$3vmb_c$: Similar to $2vmb$–$3vmb_b$.

$2vmb$–$3vmb_d$: Similar to $2vmb$–$3vmb_b$.

$2vmb$–$3vmb_e$: Similar to $2vmb$–$3vmb_b$.

$2vmb$–$3vmb_f$: Similar to $2vmb$–$3vmb_b$.

$2vmb$–$3vmb_g$: An oblique muscle, from the second ventral intersegmental membrane to the third ventral intersegmental membrane.

$2lvmb$–$3lvmb_a$: A parasternal muscle, from the second lateroventral intersegmental membrane to the third lateroventral intersegmental membrane.

$2lvmb$–$3lvmb_b$: A parasternal muscle, from the second lateroventral intersegmental membrane to the third lateroventral intersegmental membrane laterad of $2lvmb$–$3lvmb_a$.

Oblique intersegmental. There is a single, thin muscle of this type in the metathorax (Figs. 70–72, 79, 85, 91–92, 99, 112–113, 124).

t_3–$2vmb$: Anterolateral margin of the metatergum to the second ventral intersegmental membrane.

Dorsoventral. Fourteen muscles constitute this series in the metathorax (Figs. 67, 69, 72–75, 78, 80–81, 83–87, 87, 89–90, 93–94, 98, 100, 103, 106, 108–111, 115–118, 121–122, 126, 129, 131).

t_3–st_3: A thin muscle, from the anterolateral portion of the metatergum to the anterolateral portion of the metasternum.

t_3–tr_{3a}: Anterolateral margin of the metatergum to the anterodorsal margin of the metatrochanter.

t_3–tr_{3b}: Similar to t_3–tr_{3a}.

t_3–tr_{3c}: Anterolateral region of the metatergum to the anterodorsal margin of the metatrochanter.

t_3–tr_{3d}: Lateral margin of the metatergum to the laterodorsal margin of the metatrochanter.

t_3–tr_{3e}: Similar to t_3–tr_{3d}.

t_3–tr_{3f}: Similar to t_3–tr_{3d}.

t_3–tr_{3g}: Posterolateral margin of the metatergum to the laterodorsal margin of

14

the metatrochanter.

t_3–tr_{3h}: A thin muscle, from the posterolateral edge of the metatergum to the laterodorsal margin of the metatrochanter.

t_3–cx_{3a}: Anterolateral margin of the metatergum to the anterodorsal margin of the metacoxa.

t_3–cx_{3b}: Lateral region of the metatergum to the anterodorsal margin of the metacoxa.

t_3–cx_{3c}: Similar to t_3–cx_{3b}.

t_3–cx_{3d}: Similar to t_3–cx_{3b}.

t_3–cx_{3e}: A very thin muscle extending from the lateral region of the metatergum to the anterodorsal margin of the metacoxa.

Sternopleural. One muscle (Figs. 70, 72, 79, 84, 90, 92, 99, 104, 112–113, 124, 130).

$2vmb$–$3lmb$: An oblique muscle, from the second ventral intersegmental membrane to the third intersegmental membrane on its lateral portion.

Tergopleural. There are five tergopleural muscles in the metathorax (Figs. 67–68, 71–72, 74, 78–79, 84–86, 89–94, 98–99, 103, 108–109, 112–115, 117, 124, 129–131).

t_3–pl_3: A thin muscle, from the lateral region of the metatergum to the anterior edge of the metapleuron.

$2dmb$–pl_{3a}: A wide, oblique muscle, from the second dorsal intersegmental membrane to the posterior margin of the metapleuron.

$2dmb$–pl_{3b}: A short, wide, oblique muscle, from the second dorsal intersegmental membrane to the anterior portion of the metapleuron.

t_3–$3lmb_a$: Posterolateral region of the metatergum to the third lateral intersegmental membrane.

t_3–$3lmb_b$: Similar to t_3–$3lmb_a$.

Pleurocoxal. One muscle (Figs. 68, 74, 80, 84, 89, 94, 100, 103, 109, 116, 125, 129).

pl_3–cx_3: Posterior margin of the metapleuron to the anterodorsal margin of the mesocoxa.

Sternocoxal. One muscle (Figs. 68, 73, 84, 89, 94, 100, 103, 109, 116, 126, 128).

$3vmb$–cx_3: Third ventral intersegmental membrane to the anteroventral margin of the metacoxa.

Sternotrochanteral. This series in the metathorax consists of four muscles (Figs. 68, 73–74, 81, 83–84, 89, 94, 100, 103, 109, 111, 115–116, 126, 128).

$3vmb$–tr_{3a}: Third ventral intersegmental membrane to the posteroventral margin of the metatrochanter.

$3vmb$–tr_{3b}: A thin muscle, from the third ventral intersegmental membrance to the posteroventral margin of the metatrochanter.

$3vmb$–tr_{3c}: Similar to $3vmb$–tr_{3a}.

$3vmb$–tr_{3d}: A thin muscle, from the third ventral intersegmental membrane to the anterodorsal margin of the metatrochanter.

Spiracular. A small spiracle is present in the coarctate larva. There are, however, no spiracular muscles.

Table 2. Inventory of muscles of the thorax. The inventory is identical for the FG, C, and SG phases.

Prothorax	Mesothorax	Metathorax
	Dorsal longitudinal	
$1dmb\text{–}por_a$	$1dmb\text{–}2dmb_a$	$2dmb\text{–}3dmb_a$
$1dmb\text{–}por_b$	$1dmb\text{–}2dmb_b$	$2dmb\text{–}3dmb_b$
$1dmb\text{–}por_c$	$1dmb\text{–}2dmb_c$	$2dmb\text{–}3dmb_c$
$1dmb\text{–}por_d$	$1dmb\text{–}2dmb_d$	$2dmb\text{–}3dmb_d$
$1dmb\text{–}por_e$	$1dmb\text{–}2dmb_e$	$2dmb\text{–}3dmb_e$
$1dmb\text{–}por_f$	$1dmb\text{–}2dmb_f$	$2dmb\text{–}3dmb_f$
—	$1dmb\text{–}2dmb_g$	$2dmb\text{–}3dmb_g$
$t_1\text{–}por_a$	—	—
$1dmb\text{–}t_{1a}$	$t_2\text{–}2dmb_a$	$t_3\text{–}3dmb_a$
$1dmb\text{–}t_{1b}$	$t_2\text{–}2dmb_b$	$t_3\text{–}3dmb_b$
$1dmb\text{–}t_{1c}$	$t_2\text{–}2dmb_c$	$t_3\text{–}3dmb_c$
$1dmb\text{–}t_{1d}$	$t_2\text{–}2dmb_d$	$t_3\text{–}3dmb_d$
	Ventral longitudinal	
$1vmb\text{–}por_a$	$1vmb\text{–}2vmb_a$	$2vmb\text{–}3vmb_a$
$1vmb\text{–}por_b$	$1vmb\text{–}2vmb_b$	$2vmb\text{–}3vmb_b$
$1vmb\text{–}por_c$	$1vmb\text{–}2vmb_c$	$2vmb\text{–}3vmb_c$
$1vmb\text{–}por_d$	$1vmb\text{–}2vmb_d$	$2vmb\text{–}3vmb_d$
—	$1vmb\text{–}2vmb_e$	$2vmb\text{–}3vmb_e$
—	$1vmb\text{–}2vmb_f$	$2vmb\text{–}3vmb_f$
—	$1vmb\text{–}2vmb_g$	$2vmb\text{–}3vmb_g$
$1lvmb\text{–}por_a$	$1lvmb\text{–}2lvmb$	$2lvmb\text{–}3lvmb_a$
$1lvmb\text{–}por_b$	—	$2lvmb\text{–}3lvmb_b$
	Oblique intersegmental	
$st_1\text{–}por_a$	—	—
$st_1\text{–}por_b$	—	—
$st_1\text{–}por_c$	—	—
$1dmb\text{–}por_g$	—	—
$1dmb\text{–}por_h$	—	—
—	$t_2\text{–}1vmb$	$t_3\text{–}2vmb$
	Dorsoventral	
$t_1\text{–}st_1$	$t_2\text{–}st_2$	$t_3\text{–}st_3$
$t_1\text{–}vmb$	$t_2\text{–}2vmb$	—
$t_1\text{–}tr_{1a}$	$t_2\text{–}tr_{2a}$	$t_3\text{–}tr_{3a}$
$t_1\text{–}tr_{1b}$	$t_2\text{–}tr_{2b}$	$t_3\text{–}tr_{3b}$
$t_1\text{–}tr_{1c}$	$t_2\text{–}tr_{2c}$	$t_3\text{–}tr_{3c}$
$t_1\text{–}tr_{1d}$	$t_2\text{–}tr_{2d}$	$t_3\text{–}tr_{3d}$
$t_1\text{–}tr_{1e}$	$t_2\text{–}tr_{2e}$	$t_3\text{–}tr_{3e}$
$t_1\text{–}tr_{1f}$	$t_2\text{–}tr_{2f}$	$t_3\text{–}tr_{3f}$
$t_1\text{–}tr_{1g}$	$t_2\text{–}tr_{2g}$	$t_3\text{–}tr_{3g}$
$t_1\text{–}tr_{1h}$	$t_2\text{–}tr_{2h}$	$t_3\text{–}tr_{3h}$
$t_1\text{–}cx_{1a}$	$t_2\text{–}cx_{2a}$	$t_3\text{–}cx_{3a}$
$t_1\text{–}cx_{1b}$	$t_2\text{–}cx_{2b}$	$t_3\text{–}cx_{3b}$
$t_1\text{–}cx_{1c}$	$t_2\text{–}cx_{2c}$	$t_3\text{–}cx_{3c}$
$t_1\text{–}cx_{1d}$	$t_2\text{–}cx_{2d}$	$t_3\text{–}cx_{3d}$
$t_1\text{–}cx_{1e}$	$t_2\text{–}cx_{2e}$	$t_3\text{–}cx_{3e}$
	Sternopleural	
	$lvmb\text{–}2lvmb$	$2vmb\text{–}3lmb$

Prothorax	Mesothorax	Methathorax
	Tergopleural	
$t_1\text{–}por_b$	$t_2\text{–}pl_2$	$t_3\text{–}pl_3$
$t_1\text{–}pl_{1a}$	$1dmb\text{–}pl_{2a}$	$2dmb\text{–}pl_{3a}$
$t_1\text{–}pl_{1b}$	$1dmb\text{–}pl_{2b}$	$2dmb\text{–}pl_{3b}$
$t_1\text{–}pl_{1c}$	$t_2\text{–}2lmb_a$	$t_3\text{–}3lmb_a$
$t_1\text{–}pl_{1d}$	$t_2\text{–}2lmb_b$	$t_3\text{–}3lmb_b$
$t_2\text{–}pl_{1e}$	$t_2\text{–}2lmb_c$	—
	Pleurocoxal	
—	$pl_2\text{–}cx_2$	$pl_3\text{–}cx_3$
	Sternocoxal	
—	$2vmb\text{–}cx_2$	$3vmb\text{–}cx_3$
	Sternotrocanteral	
$1vmb\text{–}tr_{1a}$	$2vmb\text{–}tr_{2a}$	$3vmb\text{–}tr_{3a}$
$1vmb\text{–}tr_{1b}$	$2vmb\text{–}tr_{2b}$	$3vmb\text{–}tr_{3b}$
$1vmb\text{–}tr_{1c}$	$2vmb\text{–}tr_{2c}$	$3vmb\text{–}tr_{3c}$
—	$2vmb\text{–}tr_{2d}$	$3vmb\text{–}tr_{3d}$
	Spiracular	
$sp_1\text{–}pl_1$	—	—
$sp_1\text{–}cx_1$	—	—

17

MUSCLES OF THE ABDOMEN

The musculature of segment I of the abdomen is described in the format adopted for the head and thorax, whereas consideration of segments II–IX is restricted largely to the citation of figures and a listing of the names of muscles. The abbreviated treatment accorded the latter segments should be sufficient, however, since full parallelism in the nomenclature of the musculature of the different segments of the abdomen makes it relatively simple, in studying segments II–IX, to refer to the descriptions of serially homologous muscles in segment I. Reference to table 3 shows that differences in the pattern of musculature of segments I–VIII are minor, consisting of (1) the absence of a dorsoventral muscle (dmb–vmb) in segment I and (2) the absence of a dorsal longitudinal muscle (dmb–dmb_f) and a ventral longitudinal muscle (st–vmb) in segment VIII. The musculature of segment IX is much reduced, but the pattern reflects that of the preceding segments.

Segment I

Dorsal longitudinal. There are 13 muscles in this series (Figs. 136–141, 147–154, 156, 165–168, 170–174, 183–184). Seven of the muscles extend between the third and fourth dorsal intersegmental membranes; the remaining six run from the first abdominal tergum to the fourth dorsal intersegmental membrane.

$3dmb$–$4dmb_a$: Median portion of the third dorsal intersegmental membrane to the fourth dorsal intersegmental membrane.

$3dmb$–$4dmb_b$: Third dorsal intersegmental membrane to the fourth dorsal intersegmental membrane laterad of $3dmb$–$4dmb_a$.

$3dmb$–$4dmb_c$: Lateral portion of the third dorsal intersegmental membrane to the fourth dorsal intersegmental membrane laterad of $3dmb$–$4dmb_b$.

$3dmb$–$4dmb_d$: A paratergal muscle, well separated from the above, running from the lateral region of the third dorsal intersegmental membrane to the fourth dorsal intersegmental membrane.

$3dmb$–$4dmb_e$: A median, oblique muscle, from the third dorsal intersegmental membrane to the fourth dorsal intersegmental membrane laterad of $3dmb$–$4dmb_d$.

$3dmb$–$4dmb_f$: An oblique muscle, from the third dorsal intersegmental membrane to the fourth dorsal intersegmental membrane.

$3dmb$–$4dmb_g$: Similar to $3dmb$–$4dmb_f$; inserting laterad of $3dmb$–$4dmb_e$.

t_1–$4dmb_a$: A thin, median muscle, from the tergum of the first abdominal segment to the fourth dorsal intersegmental membrane.

19

t_I–$4dmb_b$: A thicker, oblique muscle, from the lateral portion of the first abdominal tergum to the fourth dorsal intersegmental membrane.

t_I–$4dmb_c$: Similar to t_I–$4dmb_b$.

t_I–$4dmb_d$: Similar to t_I–$4dmb_c$ but heavier.

t_I–$4dmb_e$: A wide, oblique, well separated muscle, from the anterolateral region of the first abdominal tergum to the fourth dorsal intersegmental membrane.

t_I–$4dmb_f$: A short, oblique muscle, from the posterolateral region of the first abdominal segment to the fourth dorsal intersegmental membrane.

Ventral longitudinal. Ten muscles are included in this series. Nine run between the third and fourth ventral intersegmental membranes and one extends from the sternum to the fourth intersegmental membrane (Figs. 135–140, 142–146, 149–154, 158–160, 162–164, 167–172, 176–178, 180–181).

$4vmb$–$3vmb_a$: Median portion of the fourth ventral intersegmental membrane obliquely to the third ventral intersegmental membrane.

$4vmb$–$3vmb_b$: Similar to, and laterad of, $4vmb$–$3vmb_a$.

$4vmb$–$3vmb_c$: Similar to, and laterad of, $4vmb$–$3vmb_b$.

$3vmb$–$4vmb_a$: Median part of the third ventral intersegmental membrane obliquely to the fourth ventral intersegmental membrane.

$3vmb$–$4vmb_b$: Similar to, and laterad of, $3vmb$–$4vmb_a$.

$3vmb$–$4vmb_c$: Similar to, and laterad of, $3vmb$–$4vmb_b$.

$3vmb$–$4vmb_d$: A parasternal muscle, from the third ventral intersegmental membrane to the fourth ventral intersegmental membrane.

$3vmb$–$4vmb_e$: A thin muscle, from the lateral portion of the third ventral intersegmental membrane to the fourth ventral intersegmental membrane.

$3vmb$–$4vmb_f$: Similar to $3vmb$–$4vmb_e$.

st_I–$4vmb$: A short, thin muscle, from the first abdominal sternum obliquely to the fourth ventral intersegmental membrane.

Dorsoventral. Eight muscles comprise this series (Figs. 135–140, 142, 144–147, 150–152, 154, 157, 160, 162, 164, 167, 169–170, 172, 175, 181–182). Seven run from the first abdominal tergum to the sternum, one runs from the tergum to the third ventral intersegmental membrane, and one extends between the third dorsal and ventral intersegmental membrane. The homolog of dmb–vmb in segments II–VIII is not present.

t_I–$3vmb$: Anterolateral region of the first abdominal tergum to the lateral portion of the third ventral intersegmental membrane.

t_I–st_{Ia}: A long muscle, from the anterolateral region of the first abdominal tergum to the lateral region of the first abdominal sternum.

t_I–st_{Ib}: A short, thick muscle formed by two closely appressed fascicles of equal size, running from the anterolateral margin of the first abdominal tergum to the anterolateral region of the first abdominal sternum.

t_I–st_{Ic}: A long muscle, from the lateral region of the first abdominal tergum to the lateral margin of the first abdominal sternum.

t_I–st_{Id}: Similar to t_I–st_{Ic}.

t_I–st_{Ie}: A long muscle, running from the lateral region of the first abdominal

20

tergum to the lateral region of the first abdominal sternum.

t_I–st_{If}: Very similar to t_I–st_{Ie} and lying against it.

t_I–st_{Ig}: Posterolateral region of the first abdominal tergum to the lateral margin of the first abdominal sternum.

Sternopleural. There are three muscles in this series (Figs. 135–136, 140, 144–146, 149, 154, 160, 162, 167, 172, 175, 178, 180, 182).

st_I–$3lmb$: A short muscle, from the anterolateral margin fo the first abdominal sternum to the third lateral intersegmental membrane.

st_I–$4lmb_a$: Anterolateral region of the first abdominal sternum obliquely to the fourth lateral intersegmental membrane.

st_I–$4lmb_b$: A thin muscle, from the posterolateral region of the first abdominal sternum to the fourth lateral intersegmental membrane.

Spiracular. Two thin muscles constitute this series (Figs. 149, 154, 164, 167, 169, 171–172, 175, 182).

t_I–sp_I: Lateral region of the first abdominal tergum to the first abdominal spiracle.

sp_I–st_I: First abdominal spiracle to the anterolateral region of the first abdominal sternum.

Segment II

Dorsal longitudinal. Thirteen muscles (Figs. 136–141, 147–148, 149–154, 156, 165–168, 170–174, 183–184): $4dmb$–$5dmb_a$, $4dmb$–$5dmb_b$, $4dmb$–$5dmb_c$, $4dmb$–$5dmb_d$, $4dmb$–$5dmb_e$, $4dmb$–$5dmb_f$, $4dmb$–$5dmb_g$, t_{II}–$5dmb_a$, t_{II}–$5dmb_b$, t_{II}–$5dmb_c$, t_{II}–$5dmb_d$, t_{II}–$5dmb_e$, and t_{II}–$5dmb_f$.

Ventral longitudinal. Ten muscles (Figs. 135–140, 142–144, 146, 149–154, 158–160, 162–164, 167–172, 176–178, 180–181): $5vmb$–$4vmb_a$, $5vmb$–$4vmb_b$, $5vmb$–$4vmb_c$, $4vmb$–$5vmb_a$, $4vmb$–$5vmb_b$, $4vmb$–$5vmb_c$, $4vmb$–$5vmb_d$, $4vmb$–$5vmb_e$, and $4vmb$–$5vmb_f$, and st_{II}–$5vmb$.

Dorsoventral. Nine muscles (Figs. 135–140, 142, 144, 146–147, 149–152, 154, 157, 160, 162, 164, 167–170, 172, 175, 181–183). Eight of these have obvious serial homologs in segment I: t_{II}–$4vmb$, t_{II}–st_{IIa}, t_{II}–st_{IIb}, t_{II}–st_{IIc}, t_{II}–st_{IId}, t_{II}–st_{IIe}, t_{II}–st_{IIf}, and t_{II}–st_{IIg}. In addition, there is one muscle not represented in segment I:

$4dmb$–$4vmb$: Lateral portion of the fourth dorsal intersegmental membrane at the insertion of t_I–$4dmb_c$ obliquely to the anterolateral margin of the fourth ventral intersegmental membrane.

Sternopleural. Three muscles (Figs. 135–136, 140, 142, 144, 146, 149, 154, 157, 160, 162, 167, 172, 175, 178, 180, 182): st_{II}–$4lmb$, st_{II}–$5lmb_a$, and st_{II}–$5lmb_b$.

Spiracular. Two muscles (Figs. 149, 154, 164, 167, 169, 171–172, 175, 182): t_{II}–sp_{II} and sp_{II}–st_{II}.

Segment III

Dorsal longitudinal. Thirteen muscles (Figs. 136–141, 147–154, 156, 165–168, 170–174, 183–184): $5dmb$–$6dmb_a$, $5dmb$–$6dmb_b$, $5dmb$–$6dmb_c$, $5dmb$–$6dmb_d$, $5dmb$–$6dmb_e$, $5dmb$–$6dmb_f$, $5dmb$–$6dmb_g$, t_{III}–$6dmb_a$, t_{III}–$6dmb_b$, t_{III}–$6dmb_c$, t_{III}–$6dmb_d$, t_{III}–$6dmb_e$, and t_{III}–$6dmb_f$.

Ventral longitudinal. Ten muscles (Figs. 135–140, 142–146, 149–151, 153, 158–160, 163–164, 168–172, 176–178, 180): $6vmb$–$5vmb_a$, $6vmb$–$5vmb_b$, $6vmb$–$5vmb_c$, $5vmb$–$6vmb_a$, $5vmb$–$6vmb_b$, $5vmb$–$6vmb_c$, $5vmb$–$6vmb_d$, $5vmb$–$6vmb_e$, $5vmb$–$6vmb_f$, and st_{III}–$6vmb$.

Dorsoventral. Nine muscles, as in segment II (Figs. 135–140, 142, 144–147, 149–154, 157, 160, 162, 164, 167–172, 175, 181–183): t_{III}–$5vmb$, $5dmb$–$5vmb$, t_{III}–st_{IIIa}, t_{III}–st_{IIIb}, t_{III}–st_{IIIc}, t_{III}–st_{IIId}, t_{III}–st_{IIIe}, t_{III}–st_{IIIf}, and t_{III}–st_{IIIg}.

Sternopleural. Three muscles (Figs. 135–136, 140, 142, 144–146, 149, 154, 157, 160, 162, 167, 172, 175, 178, 180, 182): st_{III}–$5lmb$, st_{III}–$6lmb_a$, and st_{III}–$6lmb_b$.

Spiracular. Two muscles (Figs. 149, 164, 167, 169, 171–172, 175, 182): t_{III}–sp_{III} and sp_{III}–st_{III}.

Segment IV

Dorsal longitudinal. Thirteen muscles (Figs. 136–140, 147–153, 156, 165–168, 170–174, 183–184): $6dmb$–$7dmb_a$, $6dmb$–$7dmb_b$, $6dmb$–$7dmb_c$, $6dmb$–$7dmb_d$, $6dmb$–$7dmb_e$, $6dmb$–$7dmb_f$, $6dmb$–$7dmb_g$, t_{IV}–$7dmb_a$, t_{IV}–$7dmb_b$, t_{IV}–$7dmb_c$, t_{IV}–$7dmb_d$, t_{IV}–$7dmb_e$, and t_{IV}–$7dmb_f$.

Ventral longitudinal. Ten muscles (Figs. 135–138, 140, 142–144, 149–154, 158–160, 162–164, 167–172, 176–178, 180): $7vmb$–$6vmb_a$, $7vmb$–$6vmb_b$, $7vmb$–$6vmb_c$, $6vmb$–$7vmb_a$, $6vmb$–$7vmb_b$, $6vmb$–$7vmb_c$, $6vmb$–$7vmb_d$, $6vmb$–$7vmb_e$, $6vmb$–$7vmb_f$, and st_{IV}–$7vmb$.

Dorsoventral. Nine muscles, as in segment II (Figs. 135–140, 142, 144, 147, 149–151, 153–154, 157, 160, 162, 164, 167–172, 175, 181–183): t_{IV}–$6vmb$, $6dmb$–$6vmb$, t_{IV}–st_{IVa}, t_{IV}–st_{IVb}, t_{IV}–st_{IVc}, t_{IV}–st_{IVd}, t_{IV}–st_{IVe}, t_{IV}–st_{IVf}, and t_{IV}–st_{IVg}.

Sternopleural. Three muscles (Figs. 135–136, 140, 144, 149, 154, 157, 160, 162, 167, 172, 175, 178, 180, 182): st_{IV}–$6lmb$, st_{IV}–$7lmb_a$, and st_{IV}–$7lmb_b$.

Spiracular. Two muscles (Figs. 149, 164, 167, 169, 171–172, 175, 182): t_{IV}–sp_{IV} and sp_{IV} and sp_{IV}–st_{IV}.

Segment V

Dorsal longitudinal. Thirteen muscles (Figs. 136–141, 147–154, 156, 165–168, 170–174, 183–184): $7dmb$–$8dmb_a$, $7dmb$–$8dmb_b$, $7dmb$–$8dmb_c$,

$7dmb$–$8dmb_d$, $7dmb$–$8dmb_e$, $7dmb$–$8dmb_f$, $7dmb$–$8dmb_g$, t_V–$8dmb_a$, t_V–$8dmb_b$, t_V–$8dmb_c$, t_V–$8dmb_d$, t_V–$8dmb_e$, and t_V–$8dmb_f$.

Ventral longitudinal. Ten muscles (Figs. 135–140, 142–146, 149–150, 152–154, 158–160, 162–164, 167–172, 176–178, 180–181): $8vmb$–$7vmb_a$, $8vmb$–$7vmb_b$, $8vmb$–$7vmb_c$, $7vmb$–$8vmb_a$, $7vmb$–$8vmb_b$, $7vmb$–$8vmb_c$, $7vmb$–$8vmb_d$, $7vmb$–$8vmb_e$, $7vmb$–$8vmb_f$, and st_V–$8vmb$.

Dorsoventral. Nine muscles, as in segment II (Figs. 135–138, 140, 142, 144–147, 149–154, 157, 160, 162, 164, 167–172, 175, 181–183): t_V–$7vmb$, $7dmb$–$7vmb$, t_V–st_{Va}, t_V–st_{Vb}, t_V–st_{Vc}, t_V–st_{Vd}, t_V–st_{Ve}, t_V–st_{Vf}, and t_V–st_{Vg}.

Sternopleural. Three muscles (Figs. 135–136, 140, 144–145, 149, 154, 160, 162, 167, 172, 178, 180, 182): st_V–$7lmb$, st_V–$8lmb_a$, and st_V–$8lmb_b$.

Spiracular. Two muscles (Figs. 149, 154, 164, 169, 171–172, 175, 182): t_{V-spV} and sp_V–st_V.

Segment VI

Dorsal longitudinal. Thirteen muscles (Figs. 136–141, 147–154, 156, 165–168, 170–174, 183–184): $8dmb$–$9dmb_a$, $8dmb$–$9dmb_b$, $8dmb$–$9dmb_c$, $8dmb$–$9dmb_d$, $8dmb$–$9dmb_e$, $8dmb$–$9dmb_f$, $8dmb$–$9dmb_g$, t_{VI}–$9dmb_a$, t_{VI}–$9dmb_b$, t_{VI}–$9dmb_c$, t_{VI}–$9dmb_d$, t_{VI}–$9dmb_d$, t_{VI}–$9dmb_f$.

Ventral longitudinal. Ten muscles (Figs. 135–140, 142–146, 149–154, 158–160, 162–164, 167–172, 176–178, 180–181): $9vmb$–$8vmb_a$, $9vmb$–$8vmb_b$, $9vmb$–$8vmb_c$, $8vmb$–$9vmb_a$, $8vmb$–$9vmb_b$, $8vmb$–$9vmb_c$, $8vmb$–$9vmb_d$, $8vmb$–$9vmb_e$, $8vmb$–$9vmb_f$, and st_{VI}–$9vmb$.

Dorsoventral. Nine muscles, as in segment II (Figs. 135–140, 142, 144–147, 149–154, 157, 160, 162, 164, 167–172, 175, 181–183): t_{VI}–$8vmb$, $8dmb$–$8vmb$, t_{VI}–st_{VIa}, t_{VI}–st_{VIb}, t_{VI}–st_{VIc}, t_{VI}–st_{VId}, t_{VI}–st_{VIe}, t_{VI}–st_{VIf}, and t_{VI}–st_{VIg}.

Sternopleural. Three muscles (Figs. 135–136, 140, 142, 144–146, 149, 154, 157, 160, 162, 167, 172, 175, 178, 180, 182): st_{VI}–$8lmb$, st_{VI}–$9lmb_a$, and st_{VI}–$9lmb_b$.

Spiracular. Two muscles (Figs. 149, 154, 164, 167, 169, 171–172, 175, 182): t_{VI}–sp_{VI} and sp_{VI}–st_{VI}.

Segment VII

Dorsal longitudinal. Thirteen muscles (Figs. 136–141, 147–154, 156, 165–168, 170–174, 183–184): $9dmb$–$10dmb_a$, $9dmb$–$10dmb_b$, $9dmb$–$10dmb_c$, $9dmb$–$10dmb_d$, $9dmb$–$10dmb_e$, $9dmb$–$10dmb_f$, $9dmb$–$10dmb_g$, t_{VII}–$10dmb_a$, t_{VII}–$10dmb_b$, t_{VII}–$10dmb_c$, t_{VII}–$10dmb_d$, t_{VII}–$10dmb_e$, and t_{VII}–$10dmb_f$.

Ventral longitudinal. Ten muscles (Figs. 135–140, 142–146, 149–154, 158–160, 162–164, 167–172, 176–178, 180–181): $10vmb$–$9vmb_a$, $10vmb$–$9vmb_b$, $10vmb$–$9vmb_c$, $9vmb$–$10vmb_a$, $9vmb$–$10vmb_b$, $9vmb$–$10vmb_c$, $9vmb$–$10vmb_d$, $9vmb$–$10vmb_e$, $9vmb$–$10vmb_f$, and st_{VII}–$10vmb$.

Dorsoventral. Nine muscles as in segment II (Figs. 135, 137–140, 142, 144–147, 149–154, 157, 160, 162, 164, 167–172, 175, 181–183):

t_{VII}–$9vmb$, $9dmb$–$9vmb$, t_{VII}–st_{VIIa}, t_{VII}–st_{VIIb}, t_{VII}–st_{VIIc}, t_{VII}–st_{VIId}, t_{VII}–st_{VIIe}, t_{VII}–st_{VIIf} and t_{VII}–st_{VIIg}.

Sternopleural. Three muscles (Figs. 135, 140, 142, 144–146, 149, 154, 157, 160, 162, 167, 172, 175, 178, 180, 182): st_{VII}–$9lmb$, st_{VII}–$10lmb_a$, and st_{VII}–$10lmb_b$.

Spiracular. Two muscles (Figs. 149, 154, 164, 167, 169, 171–172, 175, 182): t_{VII}–sp_{VII} and sp_{VII}–st_{VII}.

Segment VIII

Dorsal longitudinal. The dorsal longitudinal series in this segment includes only 12 muscles, rather than the 13 characteristic of the series in preceding abdominal segments. The missing muscle, if present, would be labelled $10dmb$–$11dmb_e$. Muscles present (Figs. 136–139, 141, 146–154, 156, 164–168, 170–174, 183–184) are: $10dmb$–$11dmb_a$, $10dmd$–$11dmb_b$, $10dmb$–$11dmb_c$, $10dmb$–$11dmb_d$, $10dmb$–$11dmb_f$, $10dmb$–$11dmb_g$, t_{VIII}–$11dmb_a$, t_{VIII}–$11dmb_b$, t_{VIII}–$11dmb_c$, t_{VIII}–$11dmb_d$, t_{VIII}–11dmb$_e$, and t_{VIII}–$11dmb_f$.

Ventral longitudinal. The number of muscles in this series is reduced from ten to nine by the absence of one of the group of muscles extending between the sternum of the segment and the eleventh ventral intersegmental membrane; if present this muscle would be labelled st_{VIII}–$11vmb$. Otherwise, the musculature is similar to segments II–VII. The muscles of the series (Figs. 136–139, 142–143, 145–147, 150–153, 158–159, 163–165, 168–171, 174, 176–177, 180–181) are: $11vmb$–$10vmb_a$, $11vmb$–$10vmb_b$, $11vmb$–$10vmb_c$, $10vmb$–$11vmb_a$, $10vmb$–$11vmb_b$, $10vmb$–$11vmb_c$, $10vmb$–$11vmb_d$, $10vmb$–$11vmb_e$, and $10vmb$–$11vmb_f$.

Dorsoventral. Nine muscles, as in segment II (Figs. 135–140, 142, 144–147, 149–154, 157, 162, 164, 167–172, 175, 181–183): t_{VIII}–$10vmb$, $10dmb$–$10vmb$, t_{VIII}–st_{VIIIa}, t_{VIII}–st_{VIIIb}, t_{VIII}–st_{VIIIc}, t_{VIII}–st_{VIIId}, t_{VIII}–st_{VIIIe}, t_{VIII}–st_{VIIIf} and t_{VIII}–st_{VIIIg}.

Sternopleural. Three muscles (Figs. 135–136, 140, 142, 144–146, 149, 153–154, 157, 160, 162, 167, 172, 175, 178, 180, 182): st_{VIII}–$10lmb$, st_{VIII}–$11lmb_a$, and st_{VIII}–$11lmb_b$.

Spiracular. Two muscles (Figs. 149, 154, 164, 167, 169, 171–172, 175, 182): t_{VIII}–sp_{VIII} and sp_{VIII}–st_{VIII}.

Segment IX

Dorsal longitudinal. There are only six muscles in this series in segment IX (Figs. 136–139, 141, 147–148, 150, 152–153, 156, 165–168, 170,

173–174, 183). This reduction results from the absence of five muscles (which would be labelled, if present, $11dmb-12dmb_b$, $11dmb-12dmb_c$, $11dmb-12dmb_e$, $11dmb-12dmb_f$, and $11dmb-12dmb_g$) in the group running between the eleventh and twelfth dorsal intersegmental membranes and two in the group running from the abdominal tergum to the twelvth dorsal intersegmental membrane ($t_{IX}-12dmb_e$ and $t_{IX}-12dmb_f$). The muscles that are present are: $11dmb-12dmb_a$, $11dmb-12dmb_d$, $t_{IX}-12dmb_a$, $t_{IX}-12dmb_b$, $t_{IX}-12dmb_c$, and $t_{IX}-12dmb_d$.

Ventral longitudinal. Of the ten muscles in this series in the other abdominal segments, only four are present in segment IX. These are: $11vmb-12vmb_a$, $11vmb-12vmb_b$, $11vmb-12vmb_c$, and $11vmb-12vmb_d$ (Figs. 136, 138, 143, 145–146, 150–153, 159, 163–164, 168–171, 177, 180–181). The missing muscles, if present, would be labelled: $12vmb-11vmb_a$, $12vmb-11vmb_b$, $12vmb-11vmb_c$, $11vmb-12vmb_e$, $11vmb-12vmb_f$, and $st_{IX}-12vmb$.

Dorsoventral. This series is reduced to four muscles (Figs. 138–140, 142, 145–147, 149, 151–152, 154, 156–157, 162, 164, 167, 169–172, 175–176, 181–182): $t_{IX}-11vmb$ (much thinner than in other segments), $t_{IX}-st_{IXc}$, $t_{IX}-st_{IXd}$, and $t_{IX}-st_{IXe}$. The missing muscles, if present, would be labelled: $11dmb-11vmb$, $t_{IX}-st_{IXa}$, $t_{IX}-st_{IXb}$, $t_{IX}-st_{IXf}$, and $t_{IX}-st_{IXg}$.

Sternopleural. There are no sternopleural muscles in segment IX.

Spiracular. There are no spiracles on segment IX.

Table 3. Inventory of muscles of the abdomen. The inventory is identical for the FG, C, and SG phases. In the general formulation of the names of the muscles employed here: $i = 3, 4, \ldots, 11$; $j = i + 1$; and $s = $ I, II, $\ldots$, IX. + means present, − absent.

Muscles (by group)	Segment								
	I	II	III	IV	V	VI	VII	VIII	IX
Dorsal longitudinal									
$(i)dmb-(j)dmb_a$	+	+	+	+	+	+	+	+	+
$(i)dmb-(j)dmb_b$	+	+	+	+	+	+	+	+	−
$(i)dmb-(j)dmb_c$	+	+	+	+	+	+	+	+	−
$(i)dmb-(j)dmb_d$	+	+	+	+	+	+	+	+	+
$(i)dmb-(j)dmb_e$	+	+	+	+	+	+	+	−	−
$(i)dmb-(j)dmb_f$	+	+	+	+	+	+	+	+	−
$(i)dmb-(j)dmb_g$	+	+	+	+	+	+	+	+	−
$t_{(s)}-(j)dmb_a$	+	+	+	+	+	+	+	+	+
$t_{(s)}-(j)dmb_b$	+	+	+	+	+	+	+	+	+
$t_{(s)}-(j)dmb_c$	+	+	+	+	+	+	+	+	+
$t_{(s)}-(j)dmb_d$	+	+	+	+	+	+	+	+	+
$t_{(s)}-(j)dmb_e$	+	+	+	+	+	+	+	+	−
$t_{(s)}-(j)dmb_f$	+	+	+	+	+	+	+	+	−
Ventral longitudinal									
$(j)vmb-(i)vmb_a$	+	+	+	+	+	+	+	+	−
$(j)vmb-(i)vmb_b$	+	+	+	+	+	+	+	+	−
$(j)vmb-(i)vmb_c$	+	+	+	+	+	+	+	+	−
$(i)vmb-(j)vmb_a$	+	+	+	+	+	+	+	+	+
$(i)vmb-(j)vmb_b$	+	+	+	+	+	+	+	+	+
$(i)vmb-(j)vmb_c$	+	+	+	+	+	+	+	+	+
$(i)vmb-(j)vmb_d$	+	+	+	+	+	+	+	+	+
$(i)vmb-(j)vmb_e$	+	+	+	+	+	+	+	+	−
$(i)vmb-(j)vmb_f$	+	+	+	+	+	+	+	+	−
$st_{(s)}-(j)vmb$	+	+	+	+	+	+	+	−	−
Dorsoventral									
$t_{(s)}-(i)vmb$	+	+	+	+	+	+	+	+	+
$(i)dmb-(i)vmb$	−	+	+	+	+	+	+	+	−
$t_{(s)}-st_{(s)a}$	+	+	+	+	+	+	+	+	−
$t_{(s)}-st_{(s)b}$	+	+	+	+	+	+	+	+	−
$t_{(s)}-st_{(s)c}$	+	+	+	+	+	+	+	+	+
$t_{(s)}-st_{(s)d}$	+	+	+	+	+	+	+	+	+
$t_{(s)}-st_{(s)e}$	+	+	+	+	+	+	+	+	+
$t_{(s)}-st_{(s)f}$	+	+	+	+	+	+	+	+	−
$t_{(s)}-st_{(s)g}$	+	+	+	+	+	+	+	+	−
Sternopleural									
$st_{(s)}-(i)lmb$	+	+	+	+	+	+	+	+	−
$st_{(s)}-(j)lmb_a$	+	+	+	+	+	+	+	+	−
$st_{(s)}-(j)lmb_b$	+	+	+	+	+	+	+	+	−
Spiracular									
$t_{(s)}-sp_{(s)}$	+	+	+	+	+	+	+	+	−
$sp_{(s)}-st_{(s)}$	+	+	+	+	+	+	+	+	−

DISCUSSION

Detailed anatomical comparison of the first grub, coarctate, and second grub phases of the blister beetle *Epicauta segmenta* has revealed remarkable continuity in the pattern of musculature during the larval stage of development. Thus, with exception of the apparent absence of a single small muscle (*St–plp*) in the head of the coarctate larva, each of the skeletal muscles present in the first grub phase is also identifiable in the coarctate and second grub phases, and neither of these phases possesses muscles not present in the first grub phase. There are, however, marked differences in the strength of development of the musculature in the three phases. In particular, the musculature undergoes profound and pervasive degeneration in the coarctate phase, only to recover to much its original massiveness and strength in the second grub phase.

First Grub Phase

In the first grub (FG) phase the muscular system consists of easily differentiated muscles, most of which are rather heavy. In the head the muscles of the antennae attach to the base of the scape and to the cranial wall instead of to the tentorium, which is somewhat reduced in this phase of development. This condition is characteristic of some other larval insects (SNODGRASS, 1935). The mandibles are each provided with two muscles, as in most pterygote insects. The maxillary musculature includes tentorio-cardinal, tentorio-stipital, occipito-palpal, and stipito-palpal elements. The galea and lacinia appear to be fused.

The labial musculature includes a pair of muscles extending from the tentorium to the prementum and a postmental muscle. A posterior frontolabral muscle is also present in the head, as in most Coleoptera (DORSEY, 1943; MAT-SUDA, 1965). The epipharynx is provided with two muscles, both taking their origin on the clypeus. Two muscles run from the frons to the pharyngeal region, but there is only one from the frons to the hypopharyngeal bar, with the usual branching. Finally, there is a branched muscle going from the occipital ridge to the pharynx.

The prothorax in the FG phase contains eleven pairs of longitudinal muscles dorsally and six pairs ventrally (table 2); two of the latter pairs attach to the lateroventral intersegmental membrane. In the mesothorax the ventral longitudinal muscles are represented by eight pairs, of which only one inserts on the corresponding lateroventral intersegmental membrane (table 2). In the metathorax there are ten pairs of ventral longitudinal muscles, three of which attach to the lateroventral membrane. Five oblique intersegmental muscles are present in the prothorax; in three of these the wider, anterior end is directed ventrad and in two dorsad. In the mesothorax and metathorax only one oblique intersegmental muscle has its anterior end directed dorsad.

27

The thoracic dorsoventral musculature includes muscles going from the tergum to the trochanter or coxa. In each of the thoracic segments one of the dorsoventral muscles (t_1–st_1) attaches to the sternal region. This condition may be unusual since SNODGRASS (1935) states that there are no tergosternal muscles represented in the prothorax of insects. MATSUDA (1970) considers reduction of the tergosternals as a general evolutionary trend in pterygote insects but recognizes the presence of such muscles in the prothorax of some insects.

In the present study no sternopleural muscles were identified in the prothorax or metathorax of the FG larva; one is present in the mesothorax (table 2). Six tergopleural muscles are represented in the prothorax and mesothorax, but only five are identifiable in the metathorax. No pleurocoxal muscles were found in the prothorax; one is present in the mesothorax and in the metathorax (table 2). The sternocoxal musculature exhibits precisely this same pattern. As far as the sternotrochanteral muscles are concerned, there are three in the prothorax and four in the mesothorax and metathorax. Thoracic spiracular muscles occur in the prothorax only. Muscles of the legs proper were not considered in this study.

The abdominal musculature in the FG phase is much the same in all segments except for a strong reduction in the number of muscles in the last segment (table 3). Thirteen pairs of dorsal longitudinal muscles occur in segments I to VII, 12 pairs in segment VIII, and only six pairs in segment IX. Ten pairs of ventral longitudinal muscles occur in segments I to VII, nine pairs in segment VIII, and only four pairs in segment IX. The first three ventral longitudinal muscles have their wider end posterior, in contrast to the condition in other muscles of this category. The wider part has been regarded as the origin in this study, as reflected in the reverse order of citation of segments in labels for muscles exhibiting this unusual condition.

Abdominal segment I has eight dorsoventral muscles, segments II to VIII nine, and segment IX only four (table 3). All abdominal segments contain three sternopleural muscles, except for the last segment, which has none.

Coarctate Phase

With the minor exception noted previously (i.e., absence of *St–plp*), the musculature of the coarctate larva is the same as that of the FG larva with respect to the distribution and arrangement of the muscles. However, within a week or so of ecdysis from the FG to C phase the entire skeletal musculature system of the larva is reduced to individual connective membranes or sheaths. These appear as thin, transluscent, silky strands which are extremely delicate and obviously nonfunctional as far as contraction is concerned. Although at least the bulk of the sarcoplasm of the muscles is lost in the process of this degeneration, it is possible to identify scattered cell nuclei within the sheaths. Whether other intracellular bodies are retained remains to be determined.

The timing of events in the degeneration of the larval musculature in the C phase has not been studied. Possibly it begins with the molt, but if so at least part of the musculature is retained in functional condition at the time of ecdysis, as demonstrated by the ability of the larva to shed its exuvia. In a matter of hours following ecdysis. however, the new cuticle, which is quite thick, becomes heavily sclerotized and tanned, and articulation of the body segments is eliminated. Thus it appears that even if the muscles were fully formed, movement of the body would be impossible. In most meloids that have been studied, including *E. segmenta*, the C phase is characterized by a deep diapause state, which may persist for several years (SELANDER and WEDDLE, 1972). The heavy, rigid integument presumably offers protection from mechanical injury and loss of water. Degeneration of the musculature, it would appear, is adaptive in minimizing the expenditure of energy for physiological maintenance during the long diapause period.

Second Grub Phase

The musculature of the second phase (SG) is well developed, fully functional, and identical in pattern and arrangement to that of the FG phase. Regeneration of the musculature begins in the C phase from two to three weeks before ecdysis to the SG phase; again, however, the precise timing of events in this developmental process has not been studied.

The individual muscles of the SG larva are generally not so wide or thick as in the FG phase and are therefore more easily dissected and identified. The weaker development of the muscles of the SG larva is associated with the fact that the SG phase is of considerably shorter duration than the FG phase (at a constant 27°C the mean duration of the SG phase is less than 11 days (SELANDER and WEDDLE, 1972)). An additional association of interest is the relatively simple behavior of the SG larva, which seems to consist largely of defense and the construction of a pupation chamber in the soil. Some recovery of muscular function is necessary, of course, for the larva to effect the ecdysis from the C phase. Finally, recovery of muscular function would appear to be dictated by the role of the muscles in controlling the shape of the body of the larva at the onset of the pupal molt. Whatever the functional elements involved, however, it is remarkable that the musculature is so thoroughly restored from the vestigial condition in the C phase. In this connection it will be most interesting to see to what extent the larval musculature is carried over to the pupal and adult stages.

Epilog

The degeneration and subsequent regeneration of the musculature described herein for *Epicauta segmenta* is the most striking ontogenetic process yet identified in hypermetamorphosis of the Meloidae. Degeneration of muscles is presumably a universal feature of the postembryonic ontogeny of endopterygote insects during metamorphosis from the larval to adult stage. In addition, it is known to occur to a limited extent in a variety of insects that do not undergo metamorphosis in the strict sense (FINLAYSON, 1975). However, the process of regeneration of previously degenerated muscles, outside the context of metamorphosis, is apparently a rare phenomenon.

In the coleopterans *Ips* (Scolytidae) and *Leptinotarsa* (Coccinellidae) there is a cyclic degeneration and regeneration of the dorsoventral indirect flight muscles of the adult (BHAKTHAN *et al.*, 1970; STEGWEE *et al.*, 1963). In *Ips* the flight muscles break down in relatively sedentary adults carrying out reproductive behavior in galleries in logs and regenerate when the beetles emerge from the logs to seek new reproductive sites. In *Leptinotarsa*, as in *Epicauta*, muscle degeneration is associated with a period of diapause.

Aside from *Epicauta*, the only other genus of insects in which a non-metamorphic cycle of degeneration and regeneration of musculature is known to occur entirely within an immature stage is the hemipteran *Rhodnius* (Reduviidae) (WIGGLESWORTH, 1956). In that genus the abdominal ventral longitudinal muscles, which play a critical role in ecdysis, grow in preparation for each molt and ecdysis, only to degenerate shortly thereafter to thin vestiges. Remnants of fibrillar structure are retained in the degenerated muscles of *Rhodnius*, as well as in those of *Ips* and *Leptinotarsa*, which is apparently not the case in *Epicauta*. Moreover, it must be emphasized that in these three genera it is only a subset of the skeletal muscles that is involved, on a selective basis, in contrast to the breakdown and reformation of the entire skeletal musculature in *Epicauta*.

We close by expressing our hope that the present work will stimulate and facilitate investigations in two quite distinct areas of biology. First, we would be pleased if it were to lead to detailed comparative studies of the musculature and other internal systems of the Meloidae—studies essential to a thorough understanding of the behavior, ecology, and evolution of this interesting and important family of beetles. Second, we feel that the ontogenetic changes that we have described offers an unparalleled opportunity for physiological studies of muscular development and maintenance. Induction and termination of diapause in the coarctate phase in *Epicauta* can be controlled by manipulation of environmental temperature and termination of diapause, at least, is easily effected by injection of the hormone ecdysone (SELANDER and WEDDLE, 1969, 1972, and unpublished). Thus, while Meloidae, as larval parasitoids, do not lend themselves to mass culture methods, cultures can be maintained in the laboratory with considerable ease and at relatively little cost.

LITERATURE CITED

ANDERSON, W. H. 1936. A comparative study of the labium of the coleopterous larvae. Smithsonian Misc. Coll. 95(13):1–29, 8 pls.

BEAUREGARD, H. 1890. Les insectes vèsicants. Paris.

BERLESE, A. 1909. Gli insetti. Milano.

BHAKTAN, N. M. G., J. H. BORDEN, and K. K. NAIR. 1970. Fine structure of degenerating and regenerating flight muscles in a bark bettle, *Ips confusus*. Jour. Cell Sci., 6:807–819.

BLAIR, J. G. 1933. Beetle larvae. Proc. and Trans. South London Ent. Nat. Hist. Soc. 1933–34:89–110.

BLOOD, R. 1935. The anatomy of *Pyrota mylabrina* (Chev.). Jour. Ent. Soc. New York, 43:1–17.

BÖVING, A. G. 1914. On the abdominal structure of certain beetle larvae of the campodeiform type. Proc. Ent. Soc. Washington, 16:55–63.

BÖVING, A. G., and F. C. CRAIGHEAD. 1930. An illustrated synopsis of the principal larval forms of the order Coleoptera. Ent. Americana, 11:1–351.

BREED, R. R. 1903. The changes which occur in the muscles of a beetle, *Thymalus marginicollis* Chev. during metamorphosis. Bull. Mus. Comp. Zool. 40:317–382, 7 pls.

BRETHÉS, J. 1917. El bicho moro (*Epicauta adspersa, Epicauta atomaria*). An. Soc. Rur. Argentina, 6:591–601.

CARPENTER, G. H., and M. C. MacDOWELL. 1972. The mouthparts of some beetle larvae (Dascillidae and Scarabaeidae), with special reference to the maxillulae and hypopharynx. Quart. Jour. Microsc. Sci. 57:373–398.

CHADWICK, L. E. 1957. The ventral intersegmental thoracic muscles of cockroaches. Smithsonian Misc. Coll. 131(11):1–30.

CHURCH, N. S., and J. G. REMPEL. 1971. The embryology of *Lytta viridana* LeConte (Coleoptera: Meloidae). Canadian Jour. Zool. 49:1563–1570.

CRAIGHEAD, E. C. 1916. The determination of the abdominal and thoracic areas of the cerambycid larvae as based on a study of the muscles. Proc. Ent. Soc. Washington, 18:129–142, pls. VI–IX.

CRAMPTON, G. C. 1918. The thoracic sclerites of immature pterygotan insects, with notes on the relationships indicated. Proc. Ent. Soc. Washington, 20:39–62, pls. 5–7.

CRAMPTON, G. C. 1921. The slerites of the head and the mouthparts of certain immature and adult insects. Ann. Ent. Soc. America, 14:65–103, pls. II–VIII.

CROME, W. 1957. Zur Morphologie und Anatomie der Larve von *Oryctes nasicornis* L. (Col. Dynastidae). Deutsches Ent. Zeitschr. (n.s.) 4:228–262.

CROS, A. 1917. *Apalus bimaculatus* L. var. *Comtei* Pic. Bull. Soc. Hist. Nat. Afrique Nord, 8:125–132.

CROS, A. 1918. Le *Meloe foveolatus* Guérin. Bull. Soc. Hist. Nat. Afrique Nord, 9:38–104. 1 pl.

CROS, A. 1919. Notes sur les larves primaires des Meloidae. Ann. Soc. Ent. France, 88:261–279.

CROS, A. 1924a. *Cerocoma Vahli* Fab. Moeurs-evolution. Mem. Soc. Hist. Nat. Afrique Nord, 15:262–292.

CROS, A. 1924b. Contribution a l'étude des espèces du genre *Sitarobrachys* Reitter et plus spécialement du *Sitarobrachys Buigasi* Escal. Bull. Soc. Sci. Nat. Maroc, 4:22–39.

CROS, A. 1926. *Zonabris impressa* Chevrolat. Bull. Soc. Sci. Nat. Maroc, 6:42–55, pls. IV–VI.

CROS, A. 1928. Essai sur la forme contractée (hypnothèque ou pseudonymphe) des larves des Meloidae. Ann. Soc. Ent. France, 97:27–58.

DAS, G. M. 1937. The musculature of the mouthparts of insect larvae. Quart. Jour. Microsc. Sci. 80:39–80, 12 pls.

DORSEY, C. K. 1943. The musculature of the labrum, labium and pharyngeal region of adult and immature Coleoptera. Smithsonian Misc. Coll. 103(7):1–42. 24 pls.

EVERLY, R. T. 1936. The alimentary tract of the margined blister beetle *Epicauta cinerea marginata*. Ohio Jour. Sci. 36:204–216, pls. I–IV.

FINLAYSON, L. H. 1975. Development and degeneration. *In* P. N. R. Usherwood (ed.), Insect muscle, pp. 75–149. New York.

GERBER, G. H., N. S. CHURCH, and J. G. REMPEL. 1971. The anatomy, histology and physiology of the reproductive systems of *Lytta nuttalli* Say (Coleoptera: Meloidae). I. The internal genitalia. Canadian Jour. Zool. 49:523–533.

31

GERBER, G. H., N. S. CHURCH, and J. G. REMPEL. 1972. The anatomy, histology and physiology of the reproductive systems of *Lytta nuttalli* Say (Coleoptera: Meloidae). II. The abdomen and external genitalia. Canadian Jour. Zool. 50:649–660.

HACHFIELD, G. 1928. Uber die Biologie und Metamorphose einer bei *Trachusa serratulae* Pz. schmarotzenden Meloide. Zeitschr. Wiss. Insektenbiol. 23:177–190.

HAYES, W. P. 1927. The immature stages and larval anatomy of *Anomala kansana* H. and McC. (Scarabaeidae. Coleop.). Ann. Ent. Soc. America, 20:193–203, pls. XI–XIII.

HAYES, W. P. 1928. The epipharynx of lamellicorn larvae (Coleop.), with a key to common genera. Ann. Ent. Soc. America, 21:282–303, pls. XV–XVII.

HAYES, W. P. 1929. Morphology, taxonomy and biology of larval Scarabaeioidea. Illinois Biol. Monogr. 12(2):1–85. 15 pls.

HORSFALL, W. R. 1943. Biology and control of common blister beetles in Arkansas. Univ. Arkansas Agr. Exp. Sta. Bull. 136, 55 pp.

HURPIN, B. 1953. Reconaissance des sexes chez les larves de Coléoptères, Scarabaeidae. Bull. Soc. Ent. France, 58:104–107.

JÖSTING, E. A. 1942. Die Innervierung des Skelettomuskelsystems des Mehlwurms (Tenebrio molitor L., Larvae). Zool. Jahrb. Abt. Anat. 67:381–460.

JUDD, W. W. 1947. The proventriculus of *Macrobasis unicolor* Kirby (Coleoptera–Melodiae [*sic*]). Ann. Ent. Soc. America, 40:518–521.

MACSWAIN, J. W. 1956. A classification of the first instar larvae of the Meloidae (Coleoptera). Univ. California Publ. Ent., 12:1–182.

MARCUZZI, G., and L. RAMPAZZO. 1960. Contributo alla conoscenza delle forme larvale dei Tenebrionidi (Col. Heteromera). Eos (Rev. Española Ent.) 36:63–117, pls. I–XIV.

MATSUDA, R. 1965. Morphology and evolution of the insect head. Ann Arbor, Michigan.

MATSUDA, R. 1970. Morphology and evolution of the insect thorax. Mem. Ent. Soc. Canada, No. 76, 431 pp.

MAYET, V. 1875. Memoire sur les moeurs et les métamorphoses d'une nouvelle espèce de Coléoptère de la famille des vèsicants le *Sitaris colletis*. Ann. Soc. Ent. France, 5:69–92, pls. III–IV.

MILLIKEN, F. B. 1921. Results of work on blister beetles in Kansas. United States Dept. Agr. Bull. 967, 26 pp.

MURRAY, F. V., and O. W. TIEGS. 1935. The metamorphosis of *Calandra oryzae*. Quart. Jour. Microsc. Sci. (n.s.) 77:405–496, pls. 23–27.

NAGATOMI, A., and K. IWATA. 1958. Biology of a Japanese blister beetle, *Epicauta gorhami* Marseul (Coleoptera, Meloidae). Mushi, 31:29–46, 1 pl.

NEWPORT, G. 1853. The natural history, anatomy and development of *Meloë*. Third memoir. The external anatomy of the larva of *Meloë* in its relation to the laws of development. Trans. Linnean Soc. London, 21:167–183, 1 pl.

PARDO ALCAIDE, A. 1953. Sobre los primeros estados evolutivos de algunos meloideos marroquíes. Tamuda, 1:87–93.

PATERSON, N. F. 1930. The bionomics and anatomy of the early stages of *Paraphaedon tumidulus* Germ. Proc. Zool. Soc. London, 3:629–676.

PERRIS, E. 1877. Larves de Coléoptères. Paris.

PETERSON, A. 1951–1962. Larvae of insects. Parts 1 and 2. Columbus, Ohio.

PRADHAN, K. S. 1949. On the head capsule, mouthparts and related muscles of the larva of the woolly bear *Anthrenus fasciatus* Herbst. (Coleoptera, Dermestidae). Rec. Indian Mus. (Calcutta) 46:73–86.

RILEY, C. V. 1877. On the larval characters and habits of the blister beetles belonging to the genera *Macrobasis* Lec. and *Epicauta* Fabr.; with remarks on other species of the family Meloidae. Trans. St. Louis Acad. Sci. 3:1–20.

ROBERTS, A. W. R. 1930. Key to the principal families of Coleoptera larvae. Bull. Ent. Res. 21:57–72.

SAXENA, O. N. 1953. Musculature of *Mylabris pustulata* Thunb. (Coleoptera). Agra Univ. Jour. Res. Sci. 2:285–307.

SAXENA, O. N. 1955. The external morphology and skeleton of *Mylabris pustulata* Thunb. (Meloidae: Coleoptera). Agra Univ. Jour. Res. Sci. 4:41–66.

SELANDER, R. B., and J. M. MATHIEU. 1964. The ontogeny of blister beetles (Coleoptera, Meloidae). I. A study of three species of the genus *Pyrota*. Ann. Ent. Soc. America, 57:711–732.

SELANDER, R. B., and R. WEDDLE. 1969. The ontogeny of blister beetles (Coleoptera, Meloidae). II. The effects of age of triungulin larvae at feeding and temperature on development in *Epicauta segmenta*. Ann. Ent. Soc. America, 62:27–39.

SELANDER, R. B., and R. WEDDLE. 1972. The ontogeny of blister beetles (Coleoptera: Meloidae). III. Diapause termination in coarctate larvae of *Epicauta segmenta*. Ann. Ent. Soc. America, 65:1–17.

SNODGRASS, R. E. 1931. Morphology of the insect abdomen. Part I. Smithsonian Misc. Coll. 85(6):1–28.

SNODGRASS, R. E. 1935. Principles of insect morphology. New York.

SPEYER, W. 1922. Die Muskulatur de Larve von *Dytiscus marginalis* L. Ein Beitrag zur Morphologie des Insektenkorpers. Zeitschr. Wiss. Zool. 119:493–492, pl. VII.

STEGWEE, D., E. C. KIMMEL, J. A. BOER, and S. HENSTRA. 1963. Hormonal control of reversible degeneration of flight muscles in the Colorado potato beetle, *Leptinotarsa decimlineata* Say (Coleoptera). Jour. Cell. Sci. 19:519–527.

STEINKE, G. 1919. Die Stigmen der Käferlarven. Arch. Naturgesch. 85(7):1–58, 2 pls.

TIEGS, O. W., and F. V. MURRAY, 1938. The embryonic development of *Calandra oryzae*. Quart. Jour. Microsc. Sci. 80:159–284, pls. 21–26.

VERHOEFF, K. 1921. Ueber vergleichende Morphologie der Mundwerkzeuge der Coleopteren Larven und Imagines. Zool. Jahrb. Abt. Syst. Oekol. Geogr. Tiere, 44:69–194.

WHITEHEAD, W. E. 1932. The morphology of the head capsule of some coleopterous larvae. Canadian Jour. Res. 6:227–252.

WIGGLESWORTH, V. B. 1956. Formation and involution of striated muscle fibres during the growth and moulting cycles of *Rhodnius prolixus* (Hemiptera). Quart. Jour. Microsc. Sci. 97:465–480.

YAKHONTOV, V. V. 1931. The pseudopupa and last larval instar of *Epicauta erythrocephala*, Pall. (Col. Meloidae). Bull. Ent. Res. 22:379–382.

33

PLATES (FIGURES 1–184)

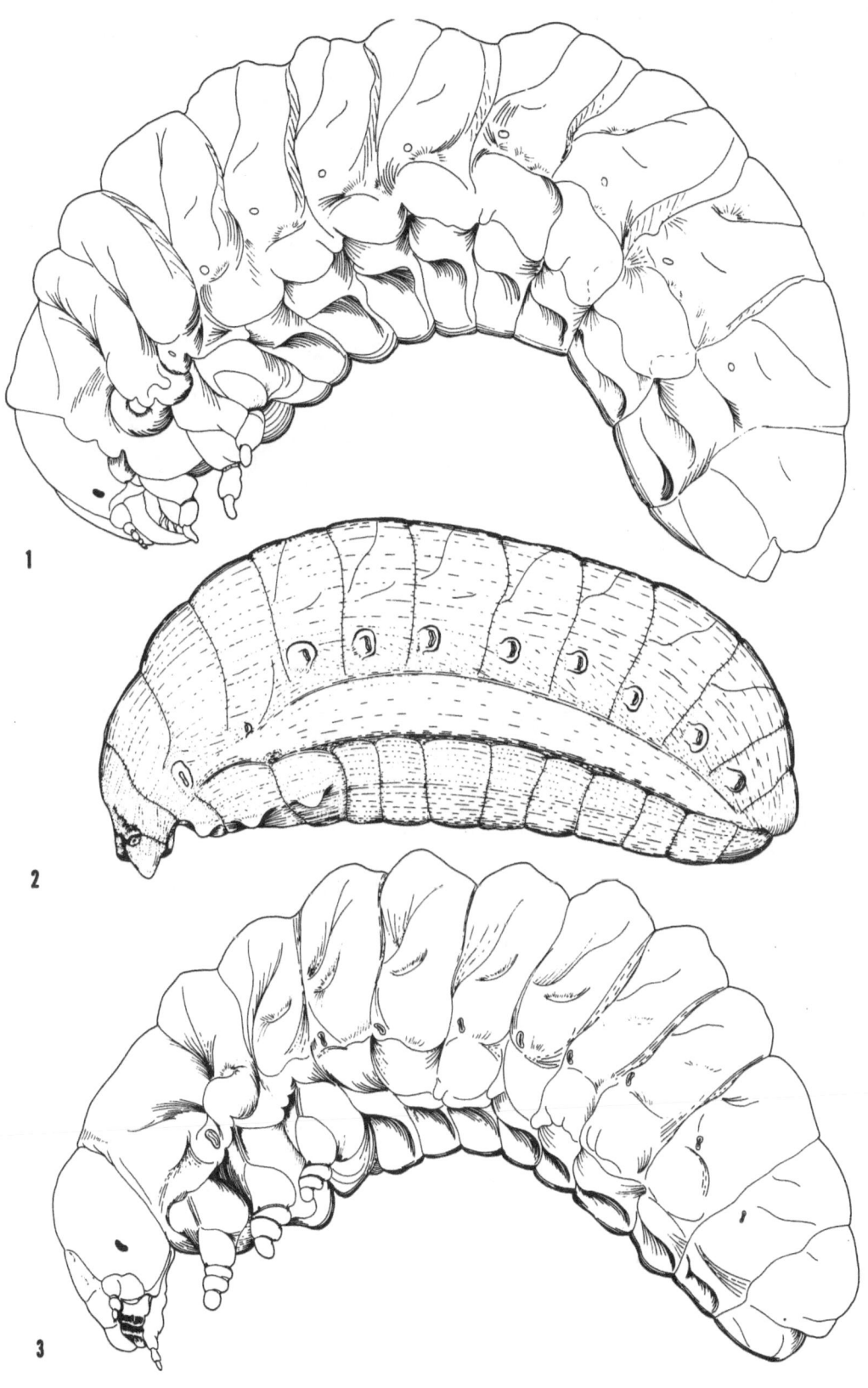

Figs. 1–3. Larva in three ontogenetic phases, lateral view. *Fig. 1.* First grub phase (FG). *Fig. 2.* Coarctate phase (C). *Fig. 3.* Second grub phase (SG).

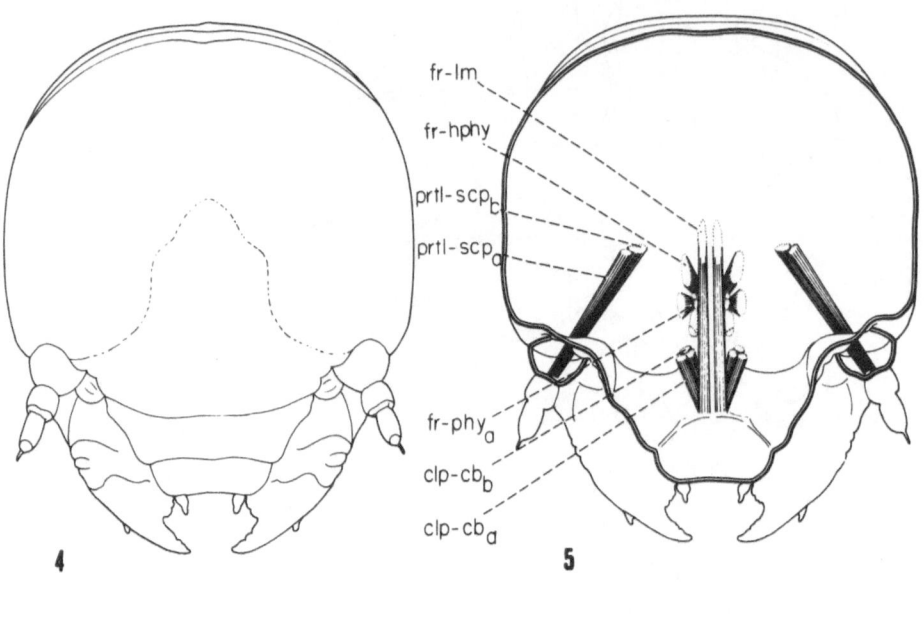

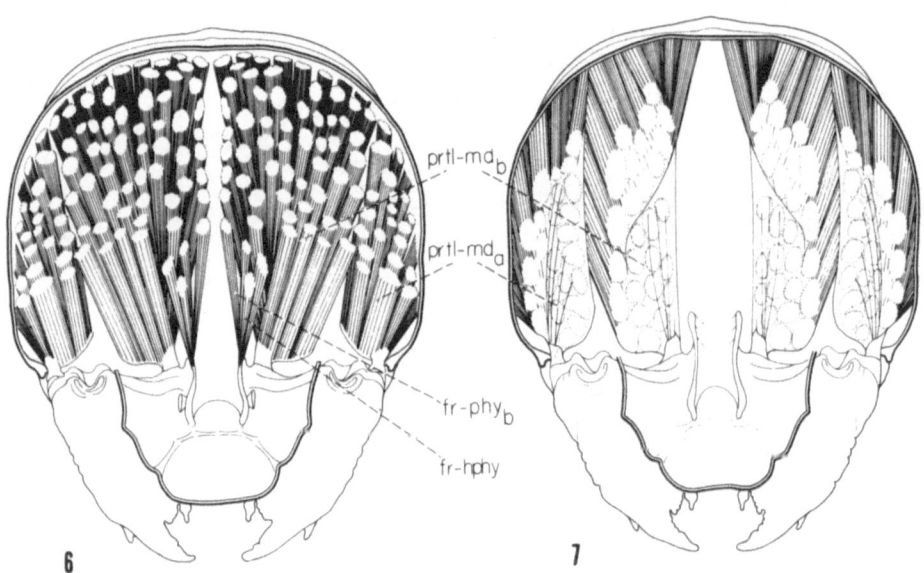

Figs. 4–7. Head of FG larva. Frontal view (see also Figs. 8–9).

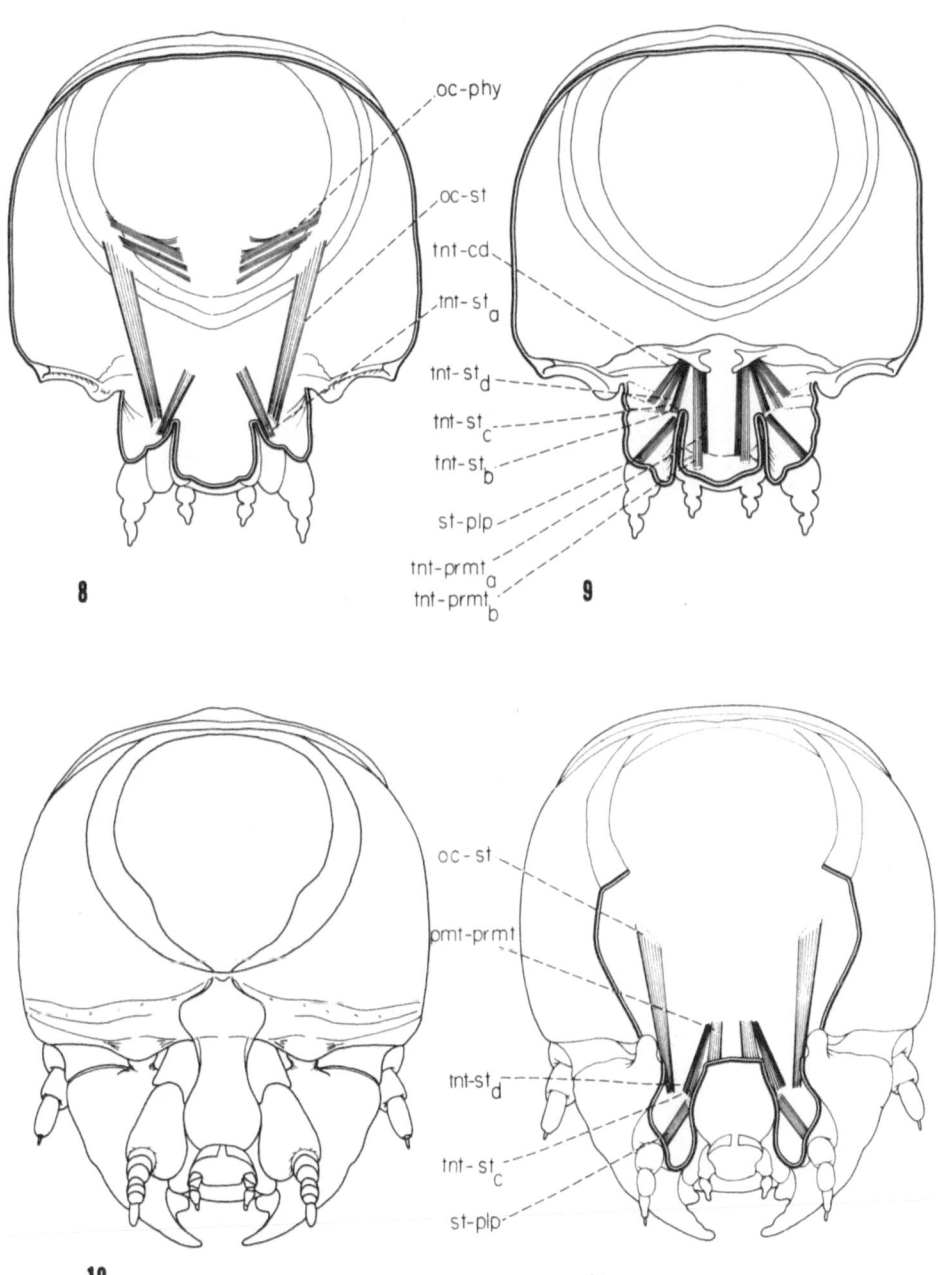

Figs. 8–11. Head of FG larva. *Figs. 8–9.* Frontal view (series concluded). *Figs. 10–11.* Posterior view (see also Figs. 12–14).

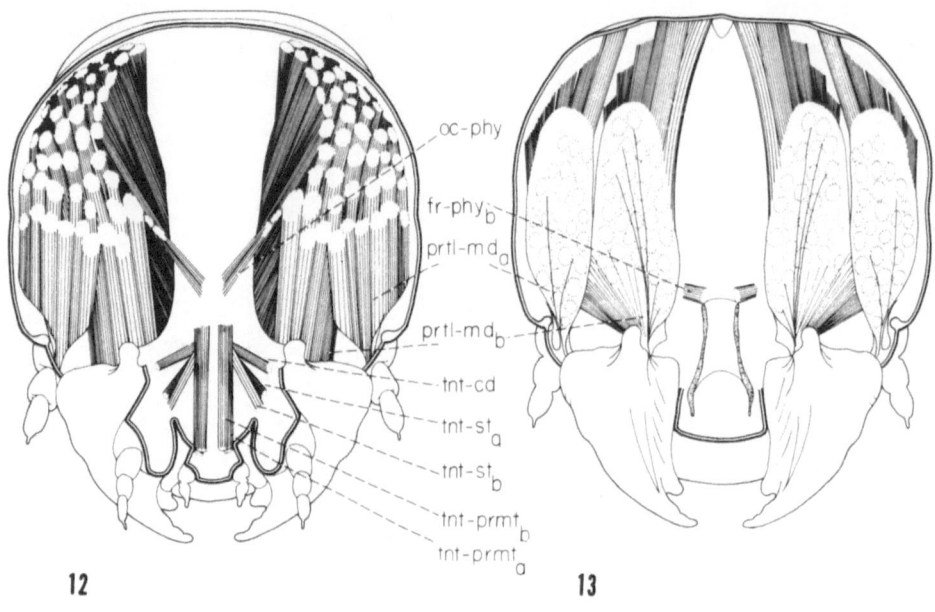

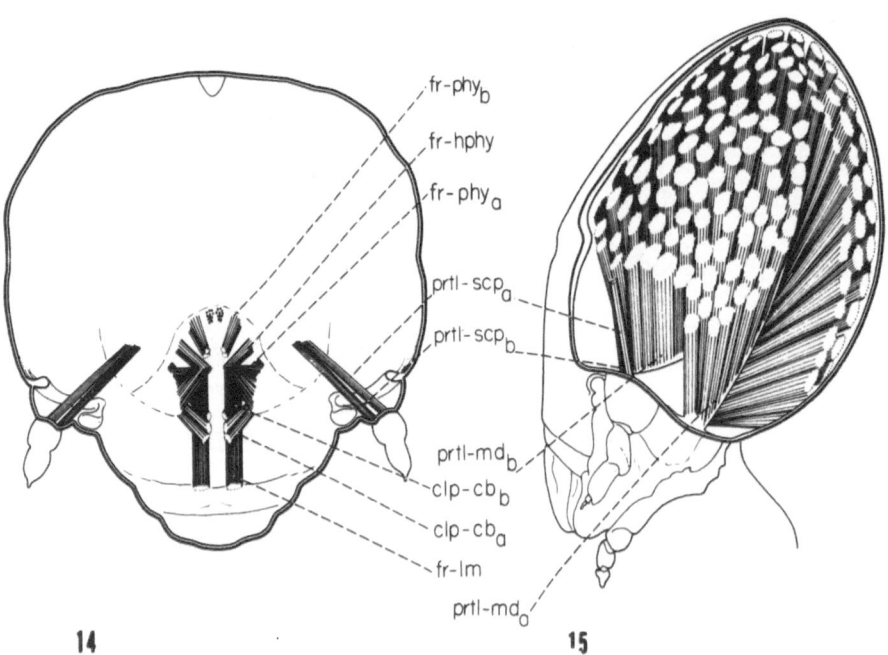

Figs. 12–15. Head of FG larva. Figs. 12–14. Posterior view (series concluded). Fig. 15. Lateral view (see also Figs. 16–20).

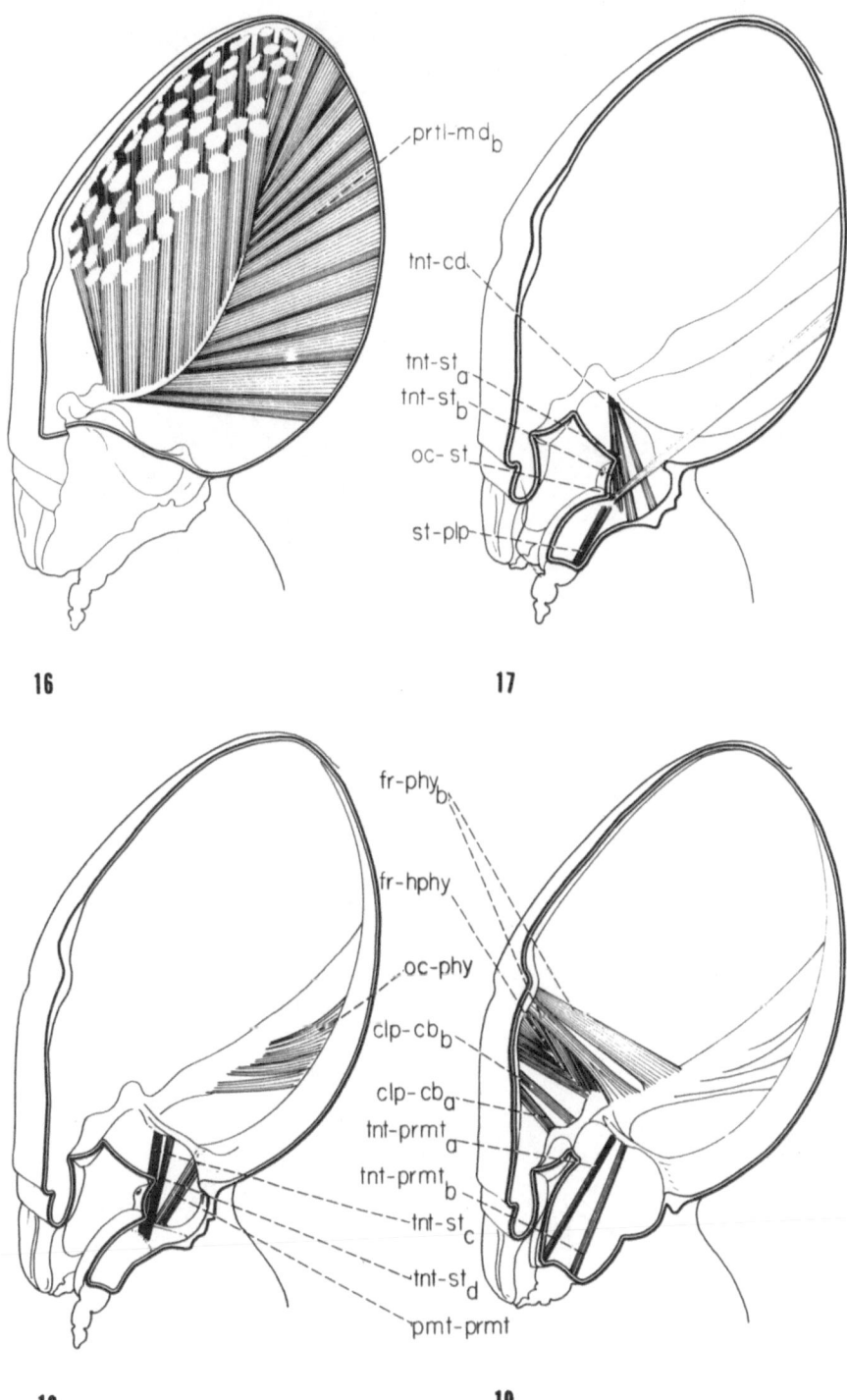

Figs. 16–19. Head of FG larva. Lateral view (series continued).

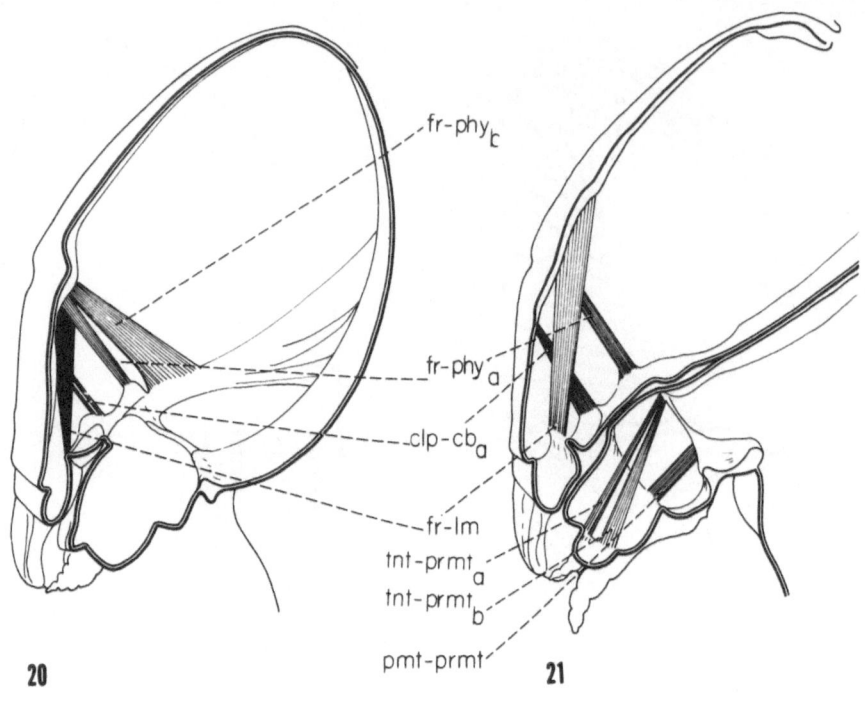

20 21

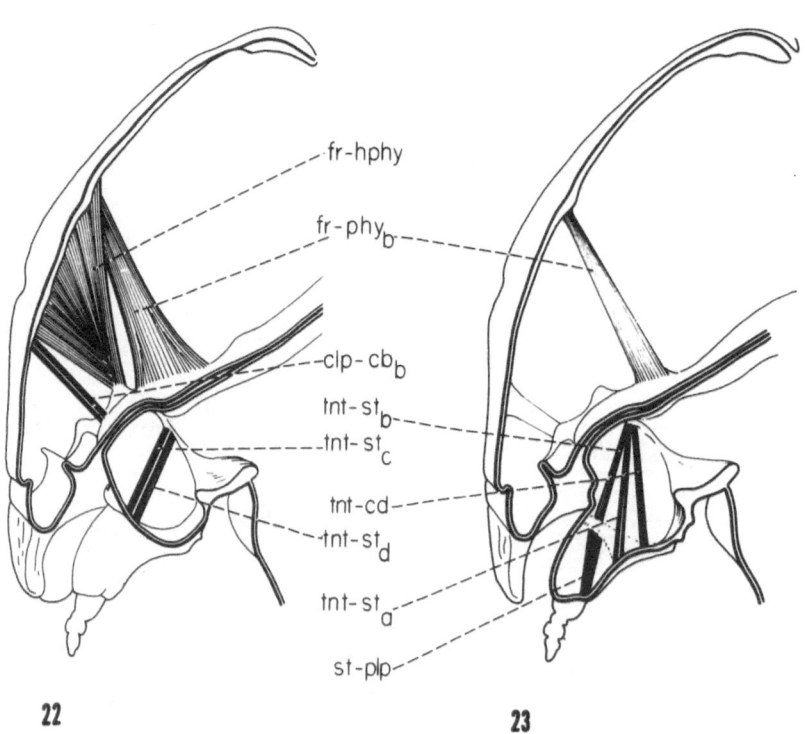

22 23

Figs. 20–23. Head of FG larva. Fig. 20. Lateral view (series concluded). Figs. 21–23. Sagittal view (see also Figs. 24–25).

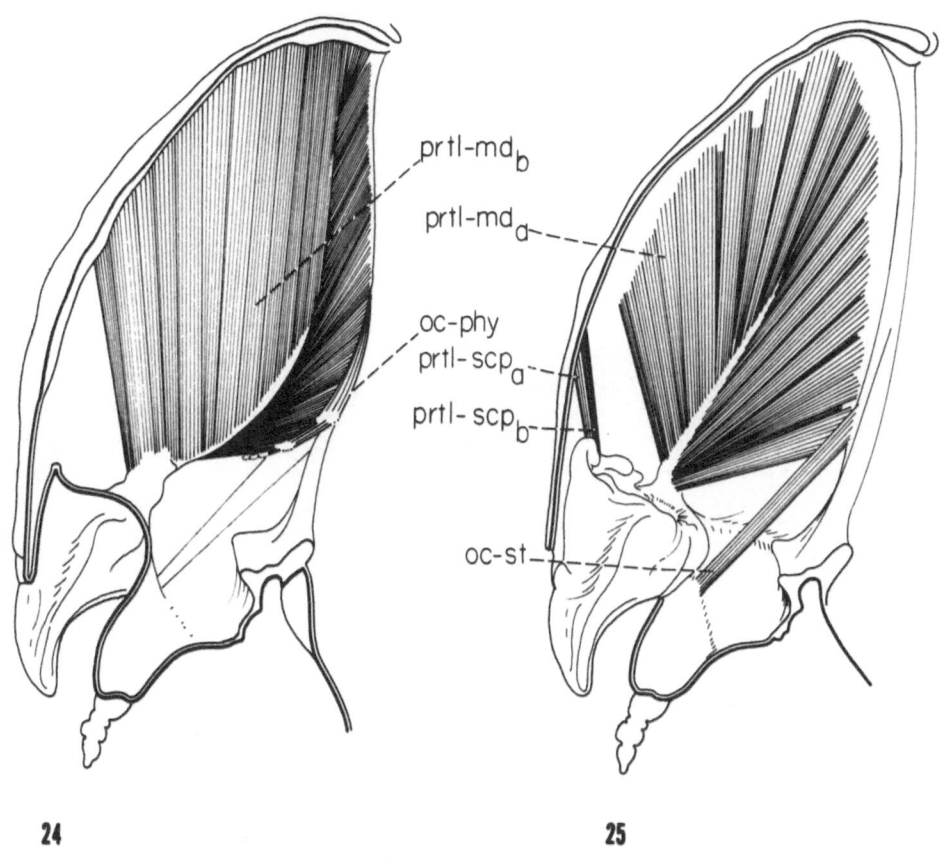

prtl-md_b

prtl-md_a

oc-phy
prtl-scp_a

prtl-scp_b

oc-st

24   **25**

Figs. 24–25. Head of FG larva. Sagittal view (series concluded).

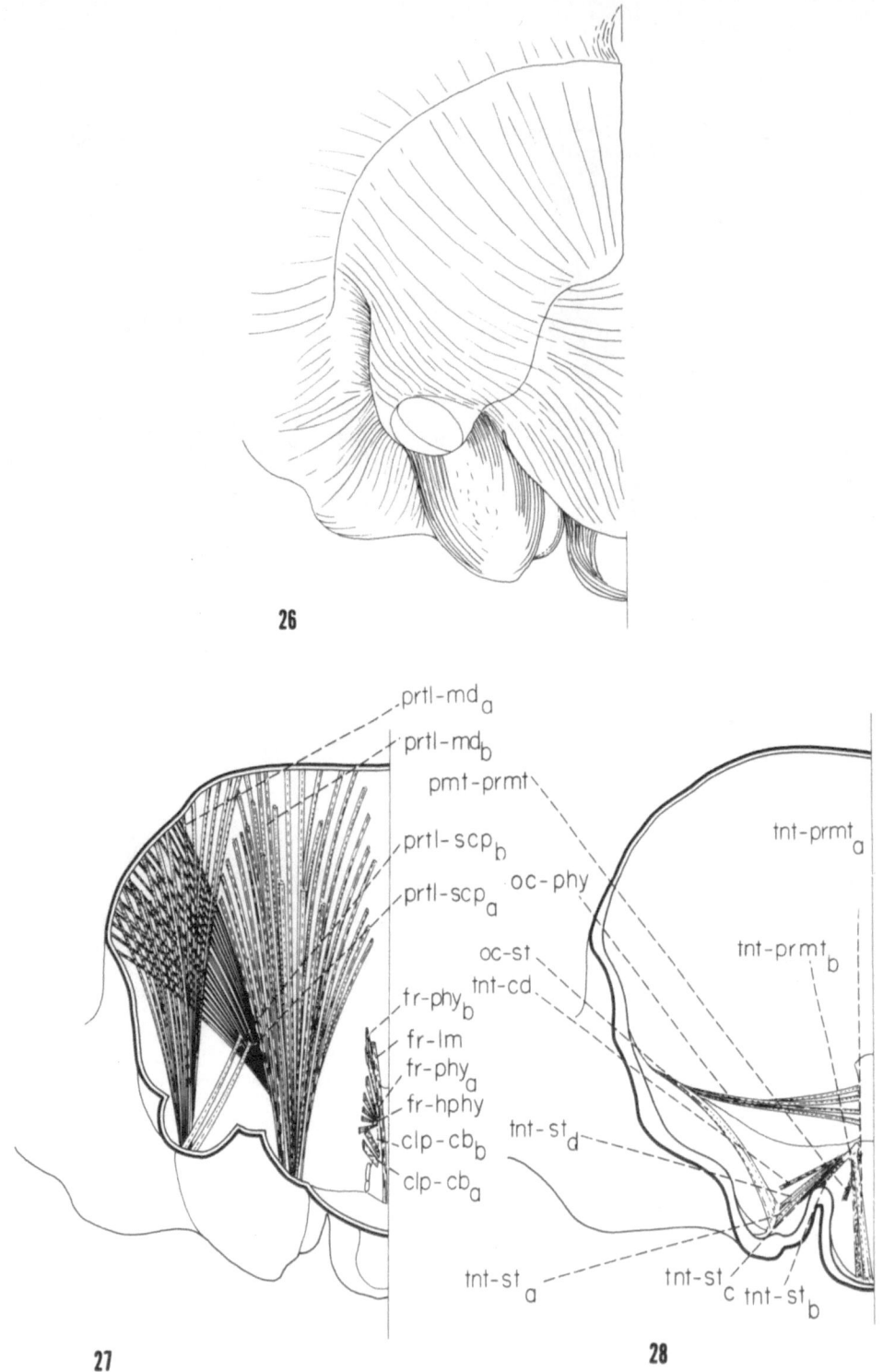

26

prtl-md$_a$
prtl-md$_b$
pmt-prmt
prtl-scp$_b$
oc-phy
prtl-scp$_a$
oc-st
tnt-cd
fr-phy$_b$
fr-lm
fr-phy$_a$
fr-hphy
clp-cb$_b$
clp-cb$_a$
tnt-st$_d$
tnt-st$_a$

tnt-prmt$_a$
tnt-prmt$_b$
tnt-st$_c$ tnt-st$_b$

27

28

Figs. 26–28. Head of C larva. Frontal view (right half).

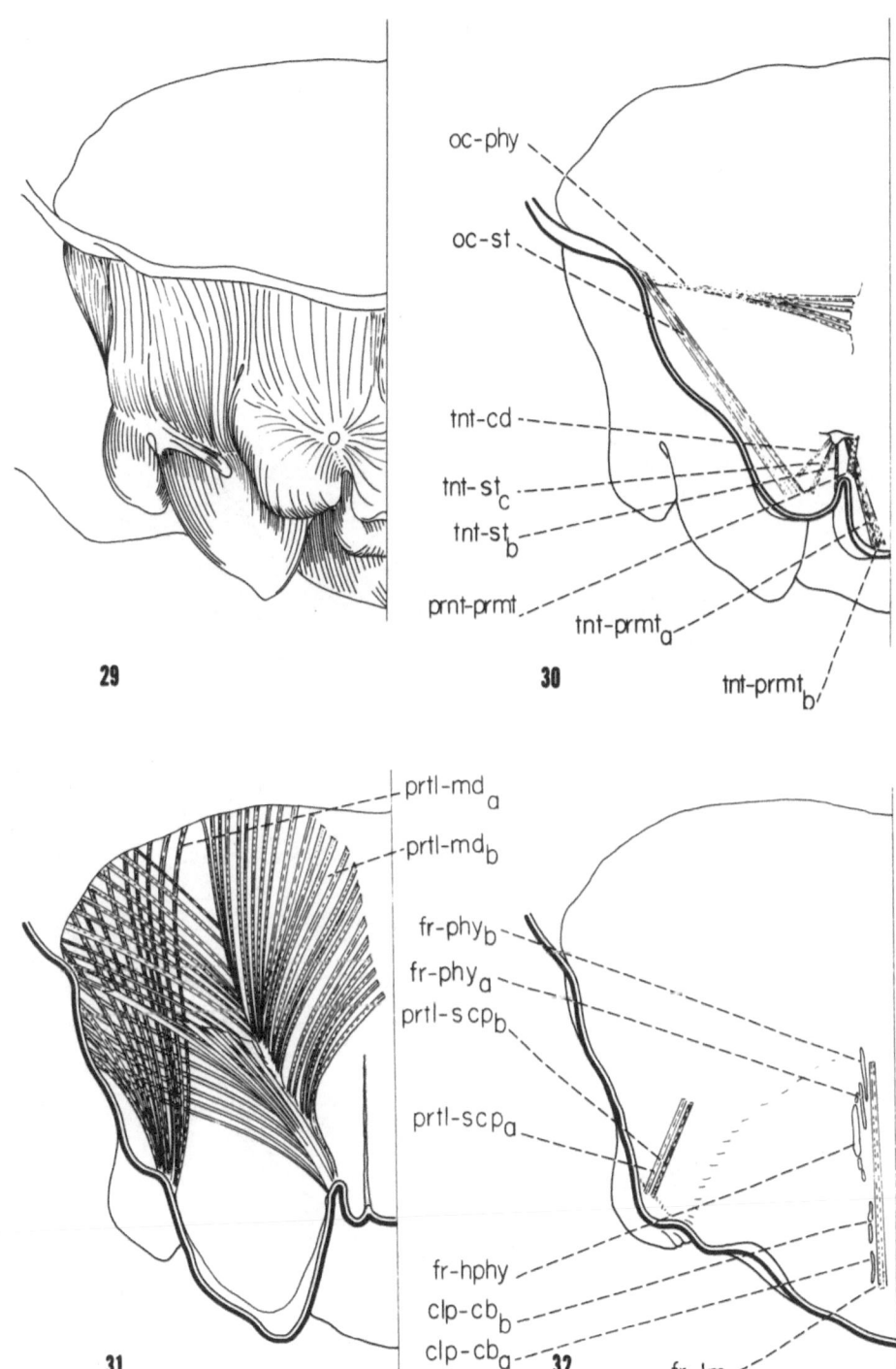

29

oc-phy
oc-st
tnt-cd
tnt-st$_c$
tnt-st$_b$
prnt-prmt
tnt-prmt$_a$

30

tnt-prmt$_b$

prtl-md$_a$
prtl-md$_b$
fr-phy$_b$
fr-phy$_a$
prtl-scp$_b$
prtl-scp$_a$
fr-hphy
clp-cb$_b$
clp-cb$_a$

31

32

fr-lm

Figs. 29–32. Head of C larva. Posterior view (left half).

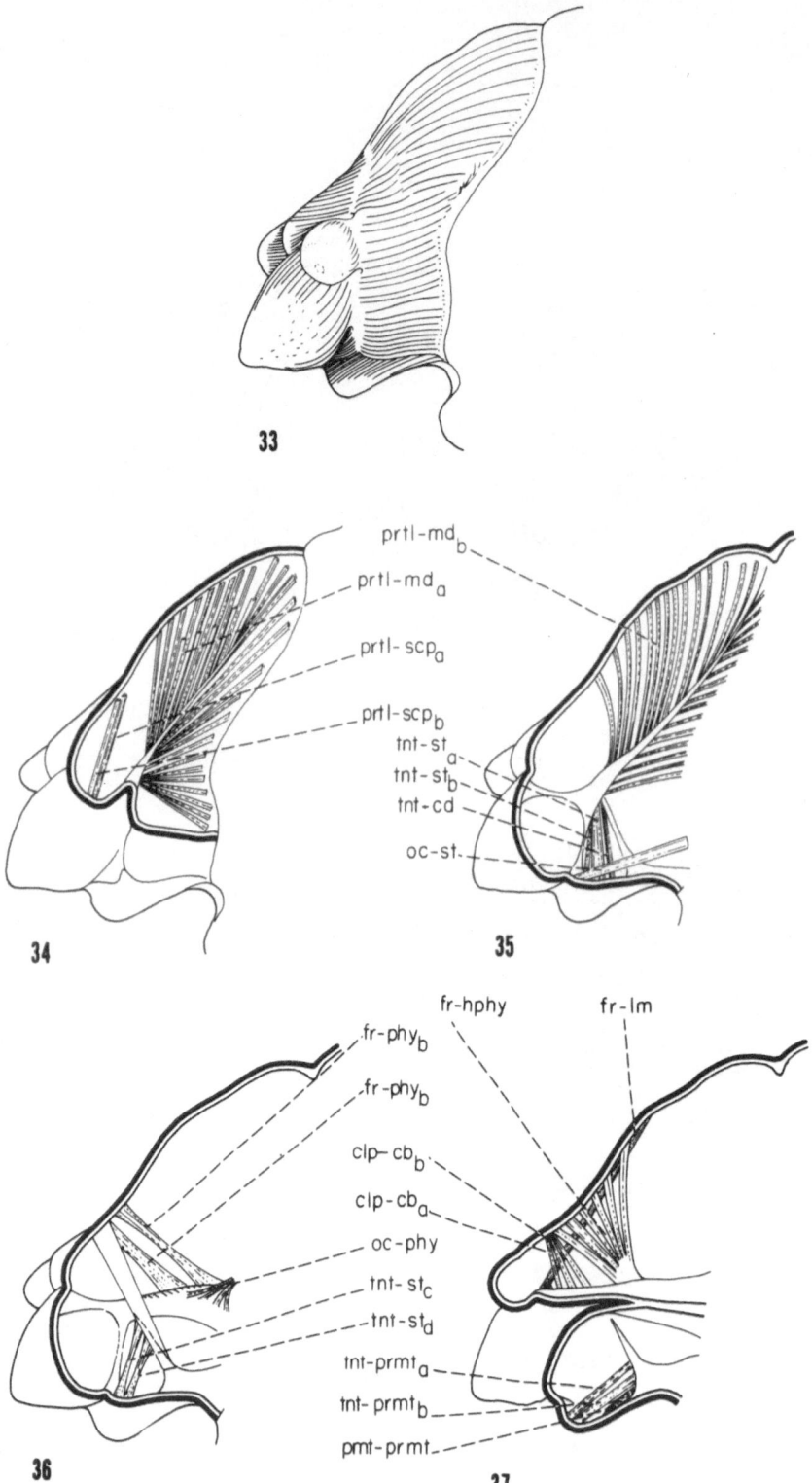

Figs. 33–37. Head of C larva. Lateral view.

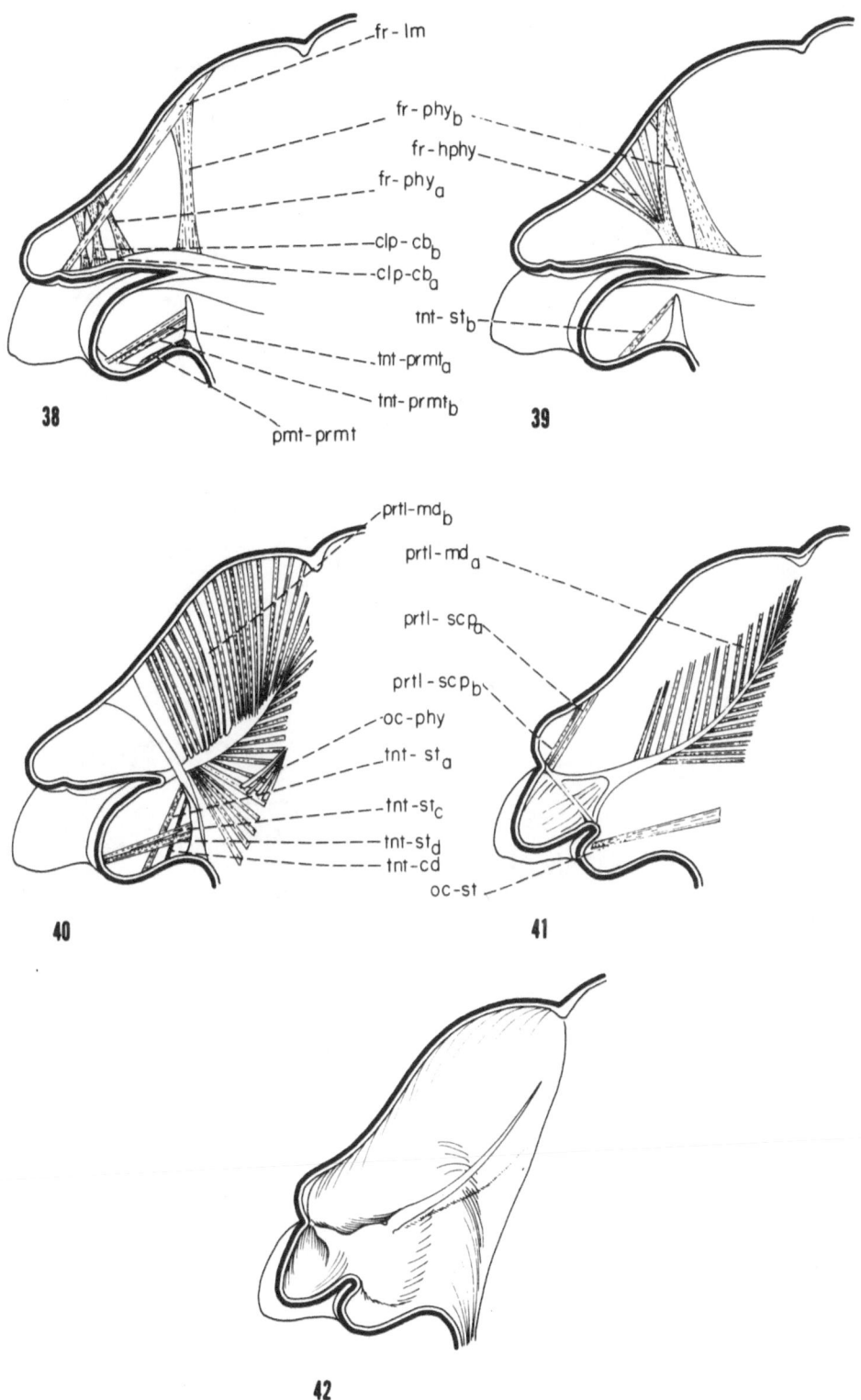

Figs. 38–42. Head of C larva. Sagittal view. All muscles removed in Fig. 42.

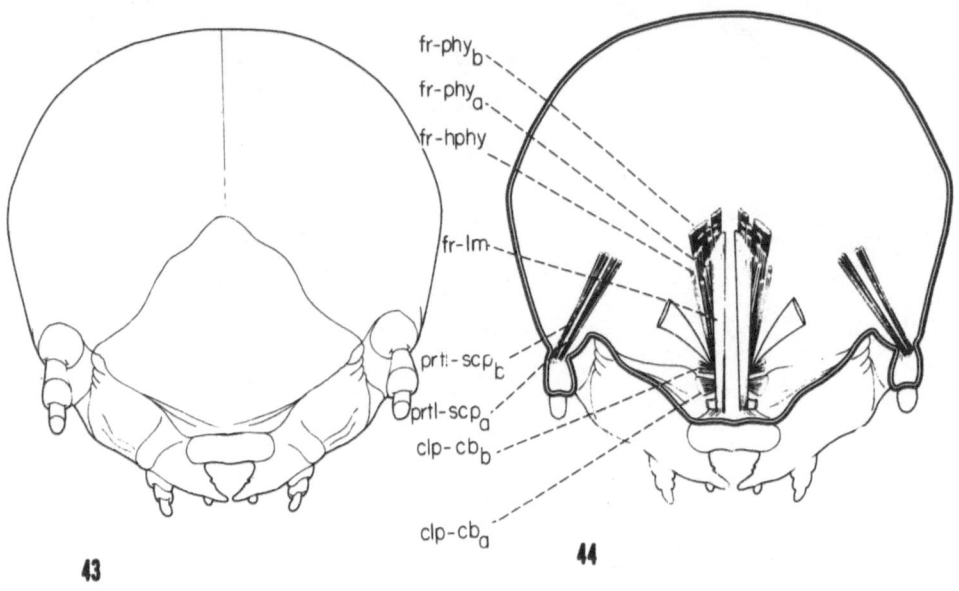

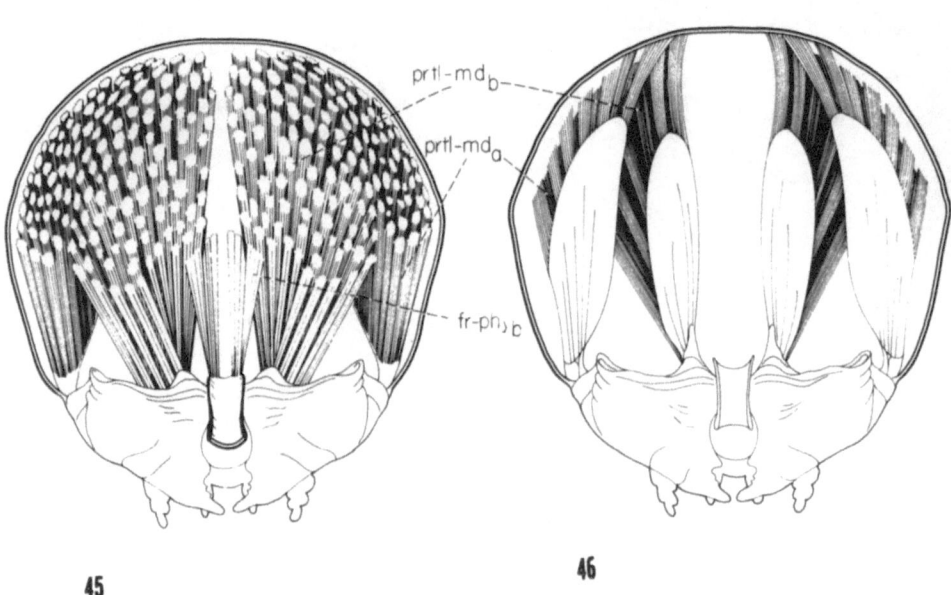

43

44

45

46

Figs. 43–46. Head of SG larva. Frontal view (see also Figs. 47–48).

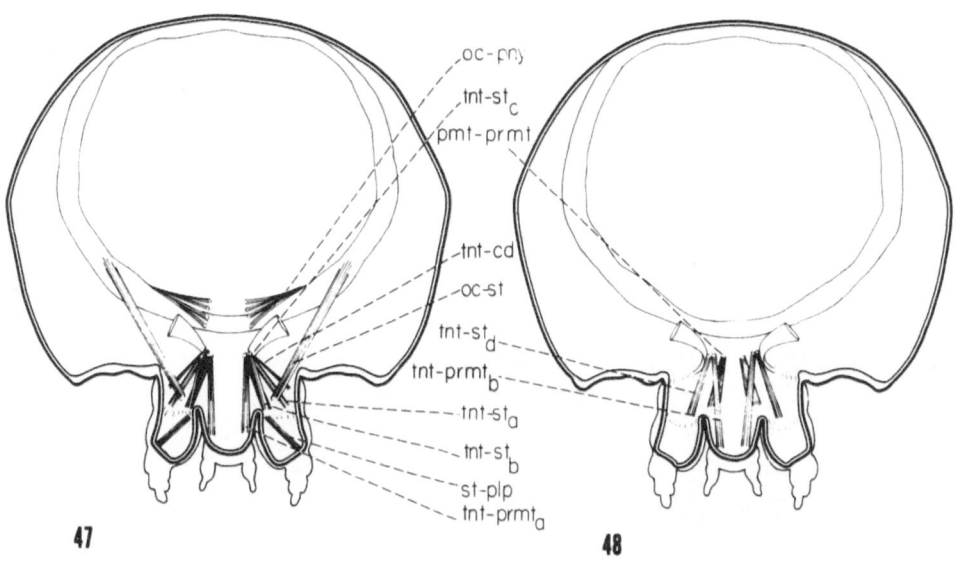

oc-phy)
tnt-st_c
pmt-prmt

tnt-cd
oc-st
tnt-st_d
tnt-prmt_b
tnt-st_a
tnt-st_b
st-plp
tnt-prmt_a

47 48

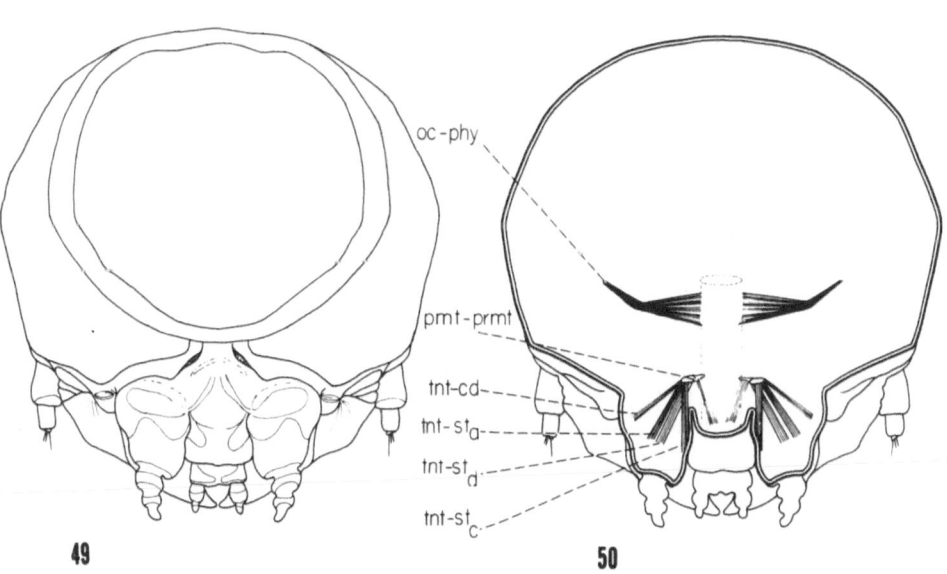

oc-phy

pmt-prmt

tnt-cd
tnt-st_a
tnt-st_d
tnt-st_c

49 50

Figs. 47–50. Head of SG larva. *Figs. 47–48.* Frontal view (series concluded). *Figs. 49–50.* Posterior view (see also Figs. 51–53).

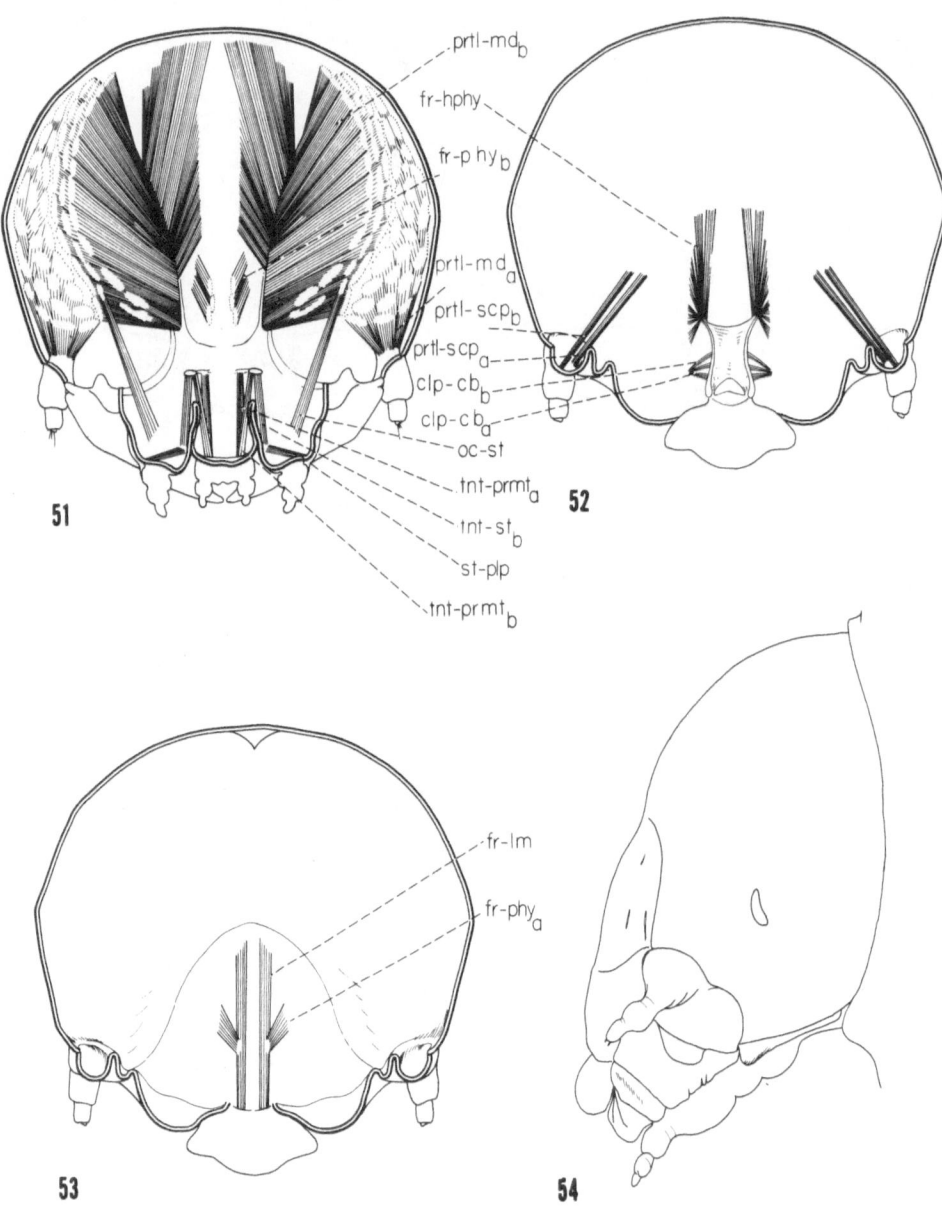

Figs. 51–54. Head of SG larva. Figs. 51–53. Posterior view (series concluded). Fig. 54. Lateral view (see also Figs. 55–59).

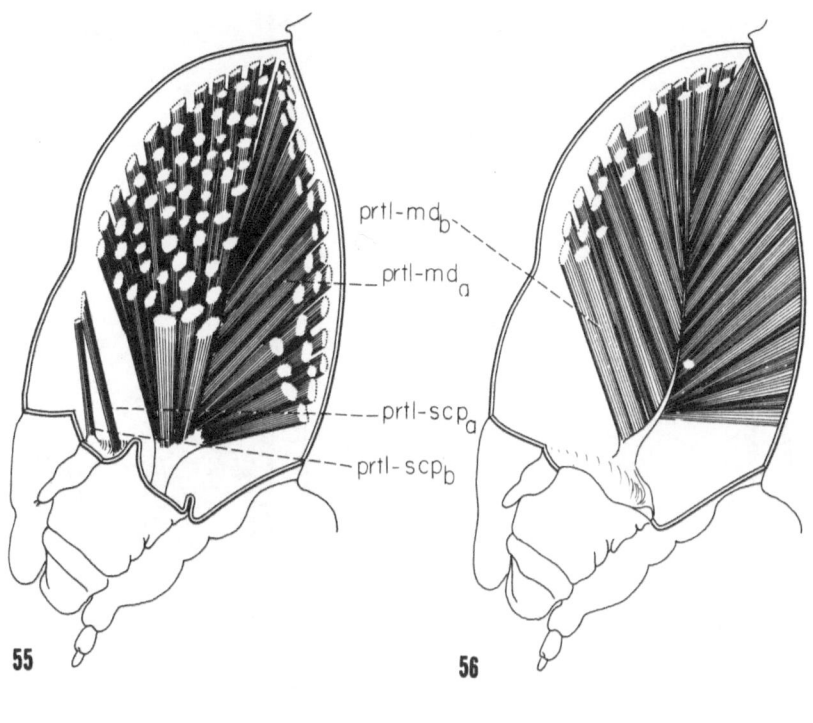

prtl-md_b

prtl-md_a

prtl-scp_a

prtl-scp_b

55 **56**

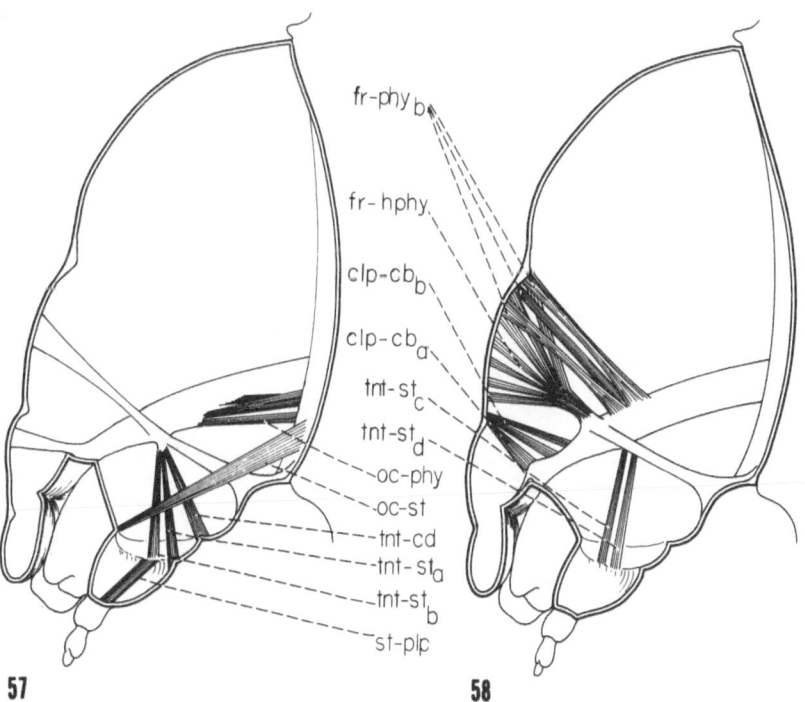

fr-phy_b

fr-hphy

clp-cb_b

clp-cb_a

tnt-st_c

tnt-st_d

oc-phy

oc-st

tnt-cd

tnt-st_a

tnt-st_b

st-plp

57 **58**

Figs. 55–58. Head of SG larva. Lateral view (series continued).

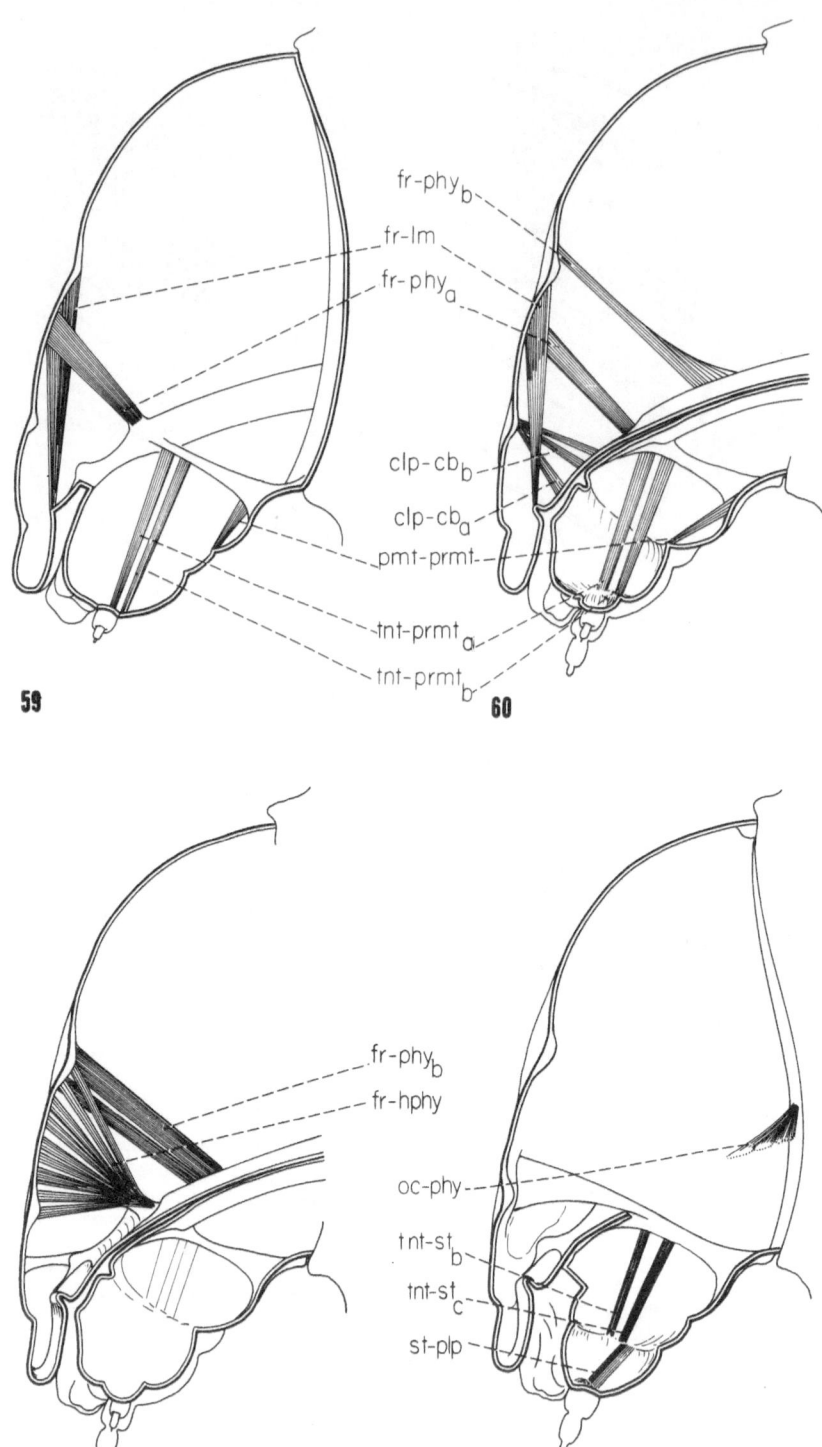

Figs. 59–62. Head of SG larva. Fig. 59. Lateral view (series concluded). Figs. 60–62. Sagittal view (see also Figs. 63–65).

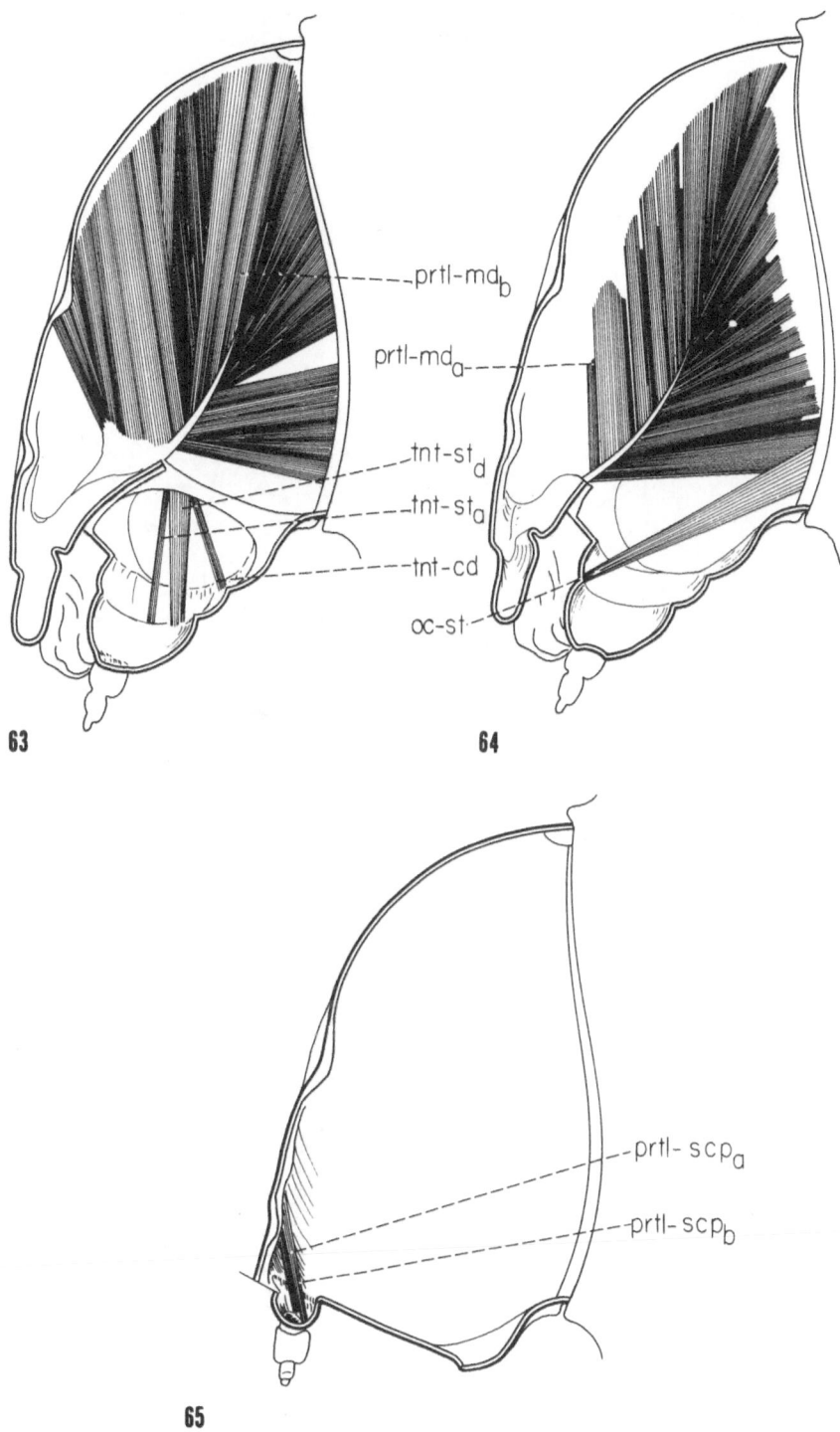

63

64

65

Figs. 63–65. Head of SG larva. Sagittal view (series concluded).

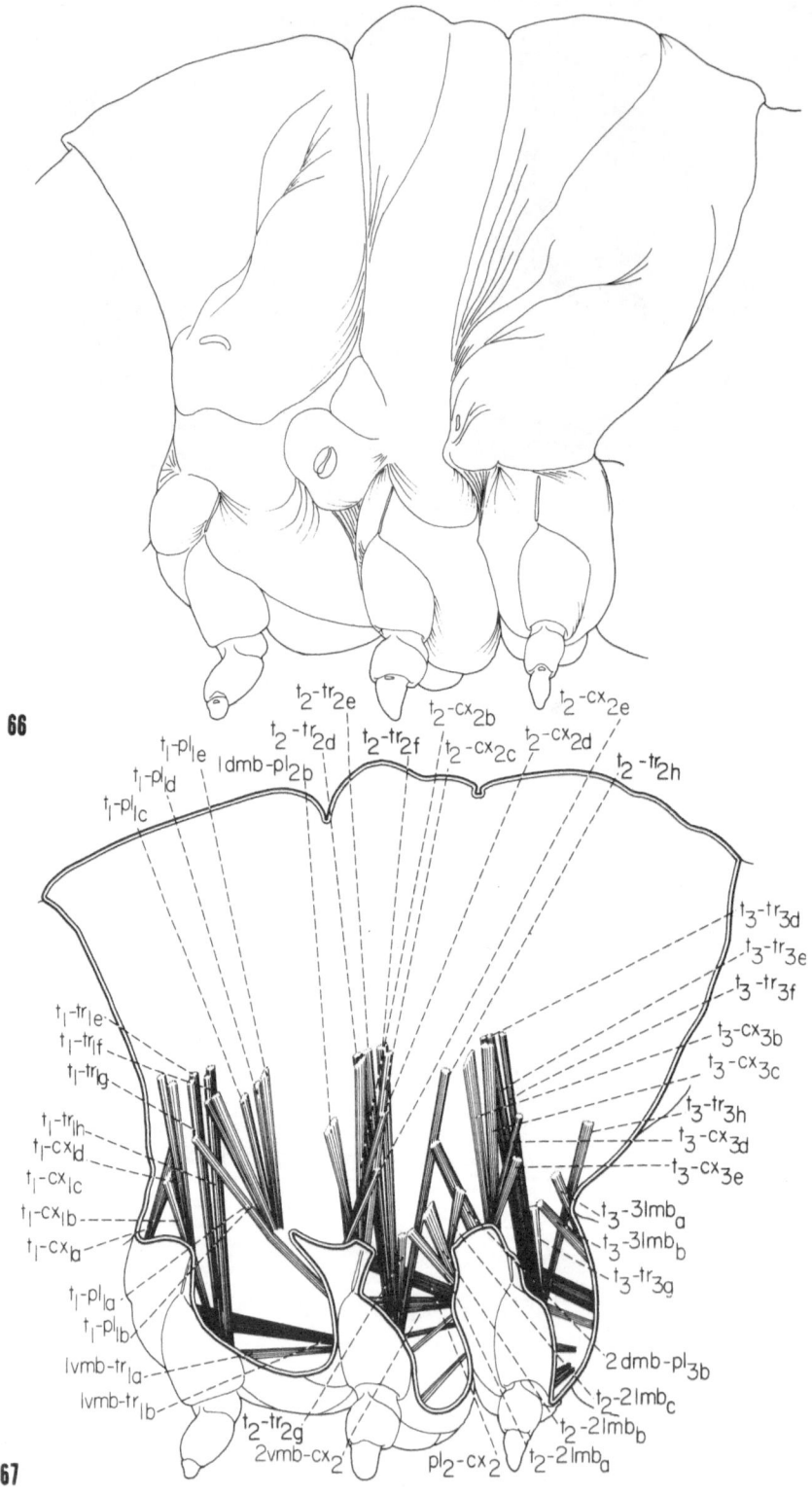

66

67

t_2-tr_{2e}
t_2-cx_{2b}
t_2-cx_{2e}
t_2-tr_{2d}
t_2-tr_{2f}
t_2-cx_{2c}
t_2-cx_{2d}
t_1-pl_{1e}
$ldmb$-pl_{2b}
t_2-tr_{2h}
t_1-pl_{1d}
t_1-pl_{1c}

t_3-tr_{3d}
t_3-tr_{3e}
t_3-tr_{3f}

t_1-tr_{1e}
t_1-tr_{1f}
t_1-tr_{1g}
t_3-cx_{3b}
t_3-cx_{3c}

t_1-tr_{1h}
t_1-cx_{1d}
t_1-cx_{1c}
t_3-tr_{3h}
t_3-cx_{3d}
t_3-cx_{3e}
t_1-cx_{1b}
t_1-cx_{1a}

t_3-$3lmb_a$
t_3-$3lmb_b$
t_1-pl_{1a}
t_3-tr_{3g}
t_1-pl_{1b}

$lvmb$-tr_{1a}
$lvmb$-tr_{1b}
2 dmb-pl_{3b}
t_2-$2lmb_c$
t_2-tr_{2g}
t_2-$2lmb_b$
$2vmb$-cx_2
pl_2-cx_2
t_2-$2lmb_a$

Figs. 66–67. Thorax of FG larva. Lateral view (see also Figs. 68–70).

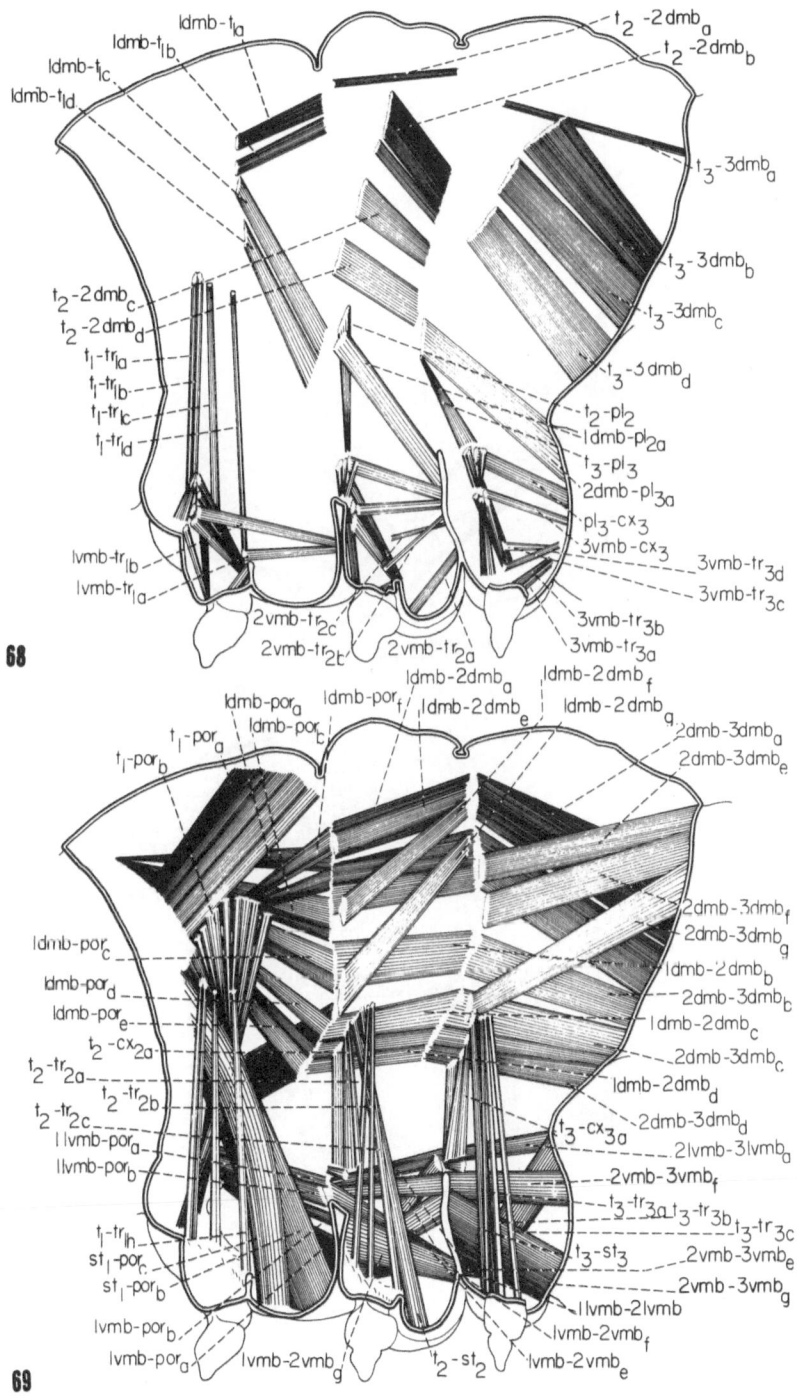

68

69

Figs. 68–69.   Thorax of FG larva. Lateral view (series continued).

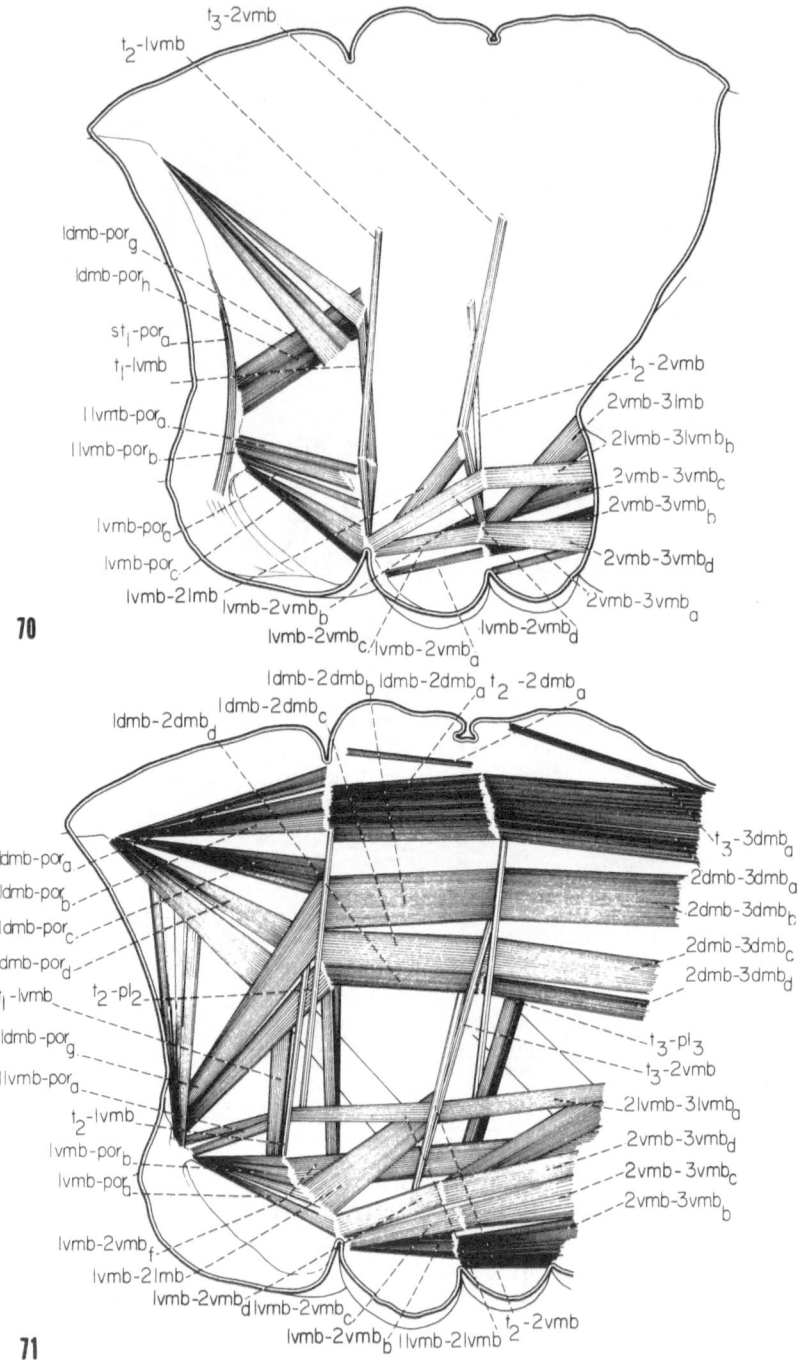

Figs. 70–71. Thorax of SG larva. *Fig. 70.* Lateral view (series concluded). *Fig. 71.* Sagittal view (see also Figs. 72–75).

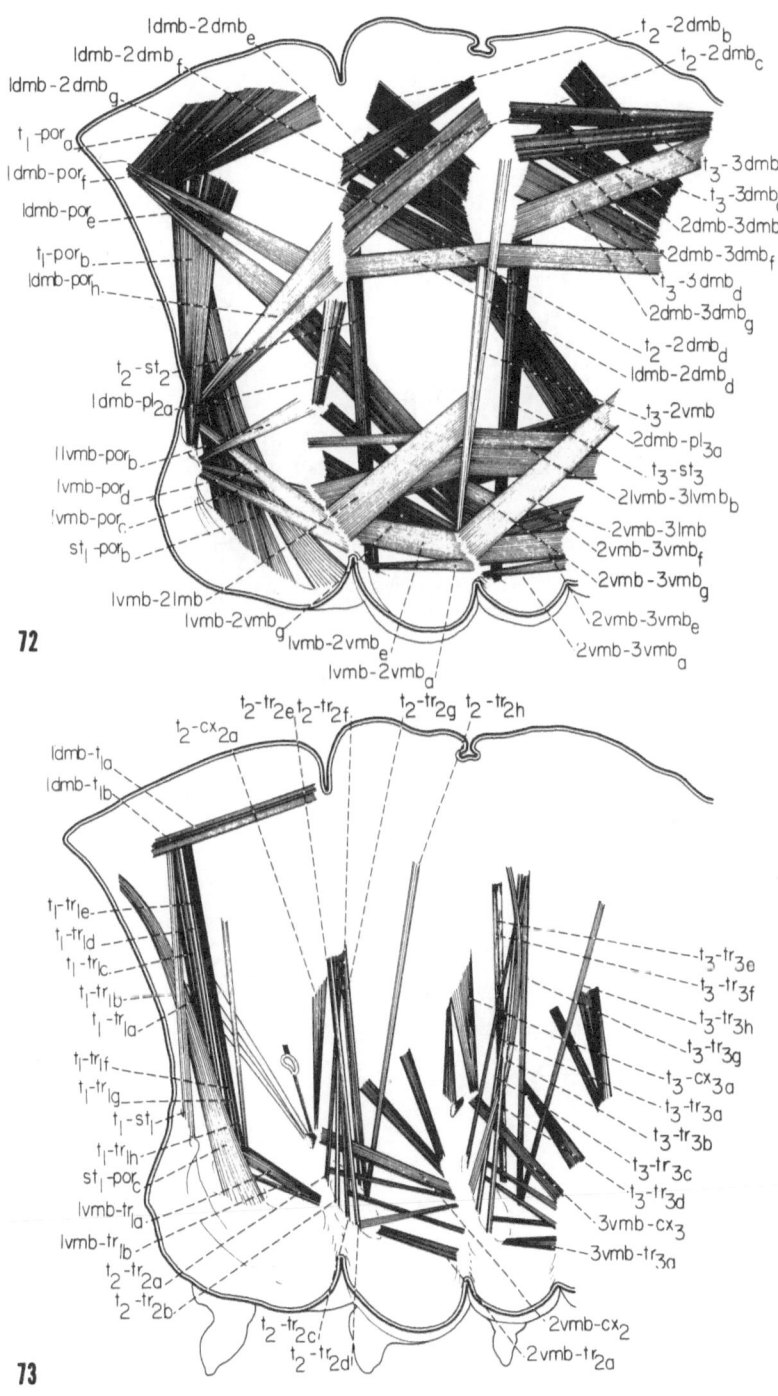

72

73

Figs. 72–73. Thorax of FG larva. Sagittal view (series continued).

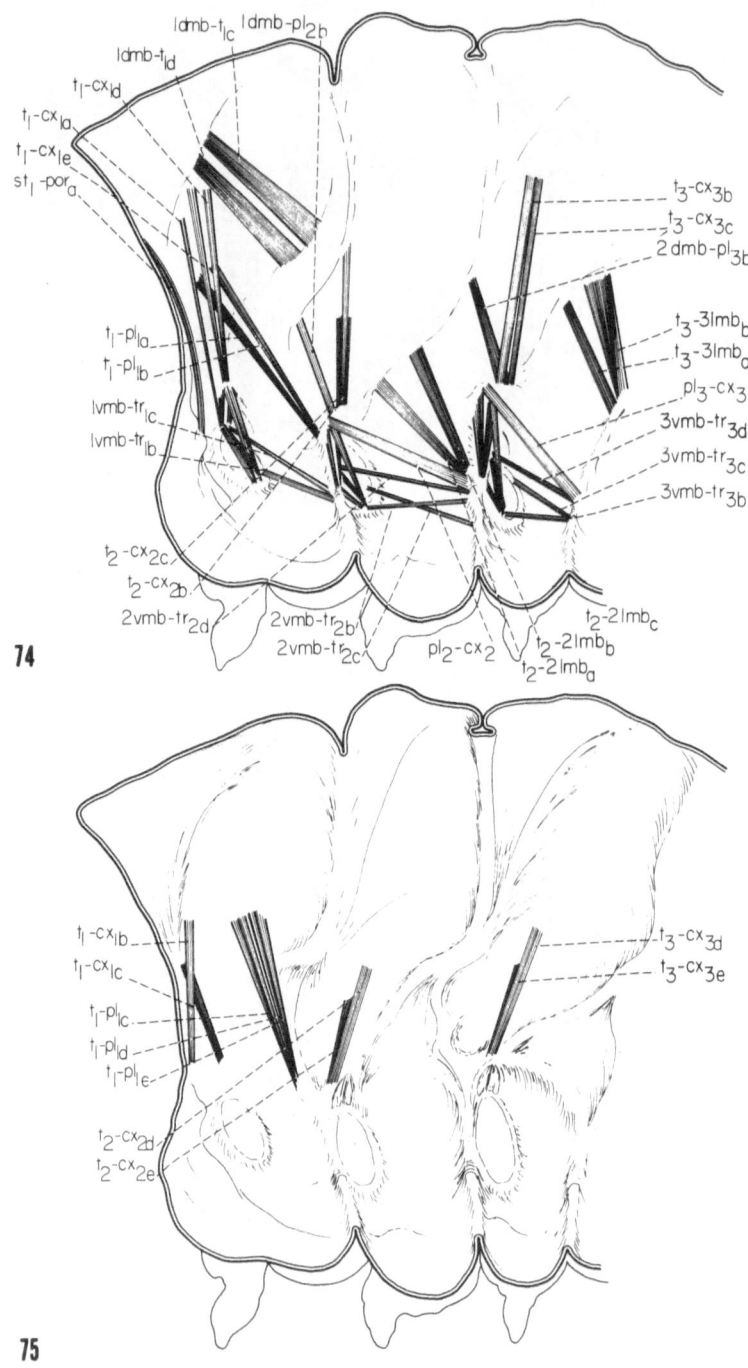

74

75

Figs. 74–75. Thorax of FG larva. Sagittal view (series concluded).

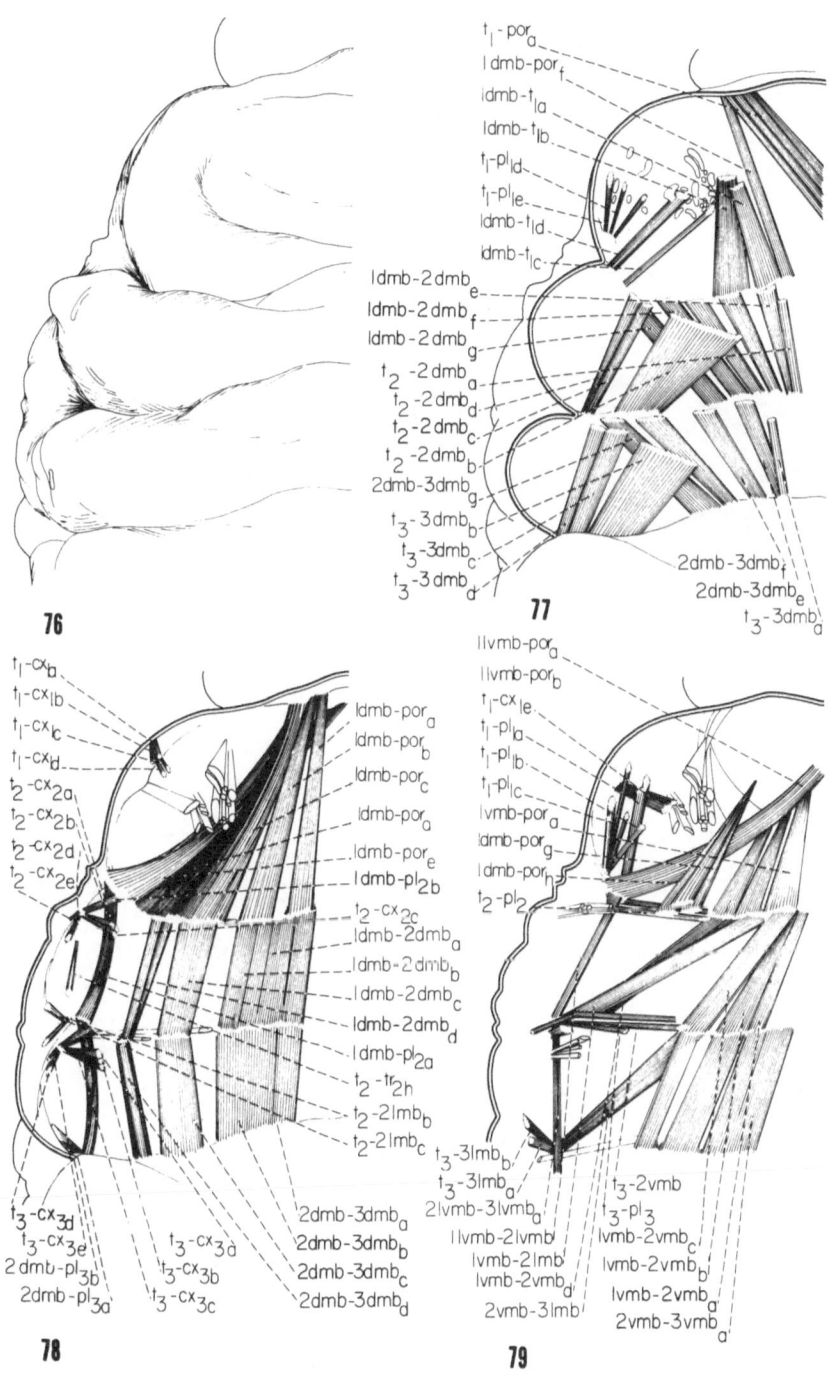

76

77

t_1-por$_a$
ldmb-por$_f$
ldmb-t_{1a}
ldmb-t_{1b}
t_1-pl$_{1d}$
t_1-pl$_{1e}$
ldmb-t_{1d}
ldmb-t_{1c}
ldmb-2dmb$_e$
ldmb-2dmb$_f$
ldmb-2dmb$_g$
t_2-2dmb$_a$
t_2-2dmb$_d$
t_2-2dmb$_c$
t_2-2dmb$_b$
2dmb-3dmb$_g$
t_3-3dmb$_b$
t_3-3dmb$_c$
t_3-3dmb$_d$

2dmb-3dmb$_f$
2dmb-3dmb$_e$
t_3-3dmb$_a$

78

t_1-cx$_{1a}$
t_1-cx$_{1b}$
t_1-cx$_{1c}$
t_1-cx$_{1d}$
t_2-cx$_{2a}$
t_2-cx$_{2b}$
t_2-cx$_{2d}$
t_2-cx$_{2e}$

ldmb-por$_a$
ldmb-por$_b$
ldmb-por$_c$
ldmb-por$_d$
ldmb-por$_e$
ldmb-pl$_{2b}$
t_2-cx$_{2c}$
ldmb-2dmb$_a$
ldmb-2dmb$_b$
ldmb-2dmb$_c$
ldmb-2dmb$_d$
ldmb-pl$_{2a}$
t_2-tr$_{2h}$
t_2-2lmb$_b$
t_2-2lmb$_c$

t_3-cx$_{3d}$
t_3-cx$_{3e}$
2dmb-pl$_{3b}$
2dmb-pl$_{3a}$
t_3-cx$_{3d}$
t_3-cx$_{3b}$
t_3-cx$_{3c}$
2dmb-3dmb$_a$
2dmb-3dmb$_b$
2dmb-3dmb$_c$
2dmb-3dmb$_d$

79

llvmb-por$_a$
llvmb-por$_b$
t_1-cx$_{1e}$
t_1-pl$_{1a}$
t_1-pl$_{1b}$
t_1-pl$_{1c}$
lvmb-por$_a$
ldmb-por$_g$
ldmb-por$_h$
t_2-pl$_2$

t_3-3lmb$_b$
t_3-3lmb$_a$
2lvmb-3lvmb$_a$
llvmb-2lvmb
lvmb-2lmb
lvmb-2vmb$_d$
2vmb-3lmb

t_3-2vmb
t_3-pl$_3$
lvmb-2vmb$_c$
lvmb-2vmb$_b$
lvmb-2vmb$_a$
2vmb-3vmb$_a$

Figs. 76–77. Thorax of FG larva. Dorsal view (left half) (see also Figs. 80–81).

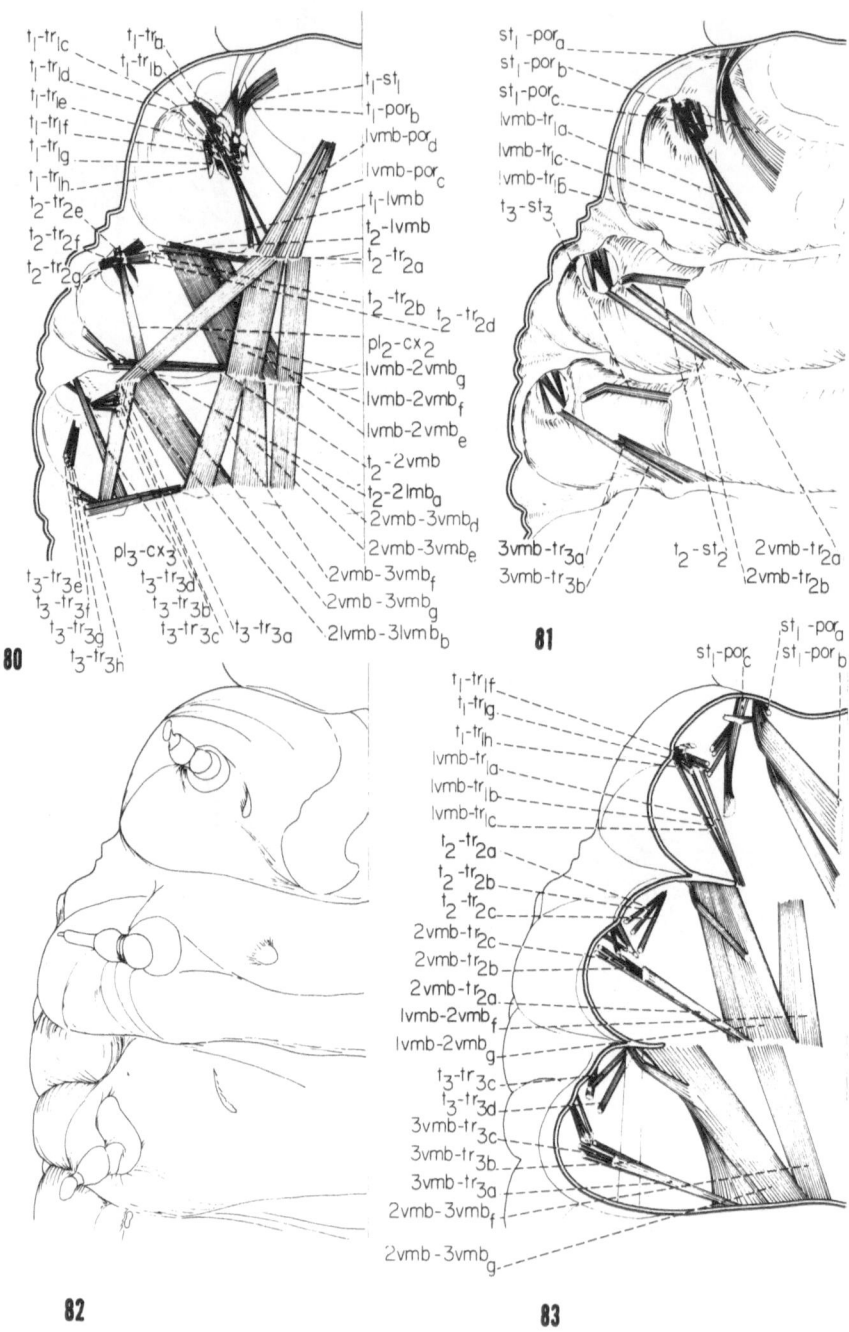

Figs. 80–83. Thorax of FG larva. *Figs. 80–81.* Dorsal view (left half) (series concluded). *Figs. 82–83.* Ventral view (right half) (see also Figs. 84–87).

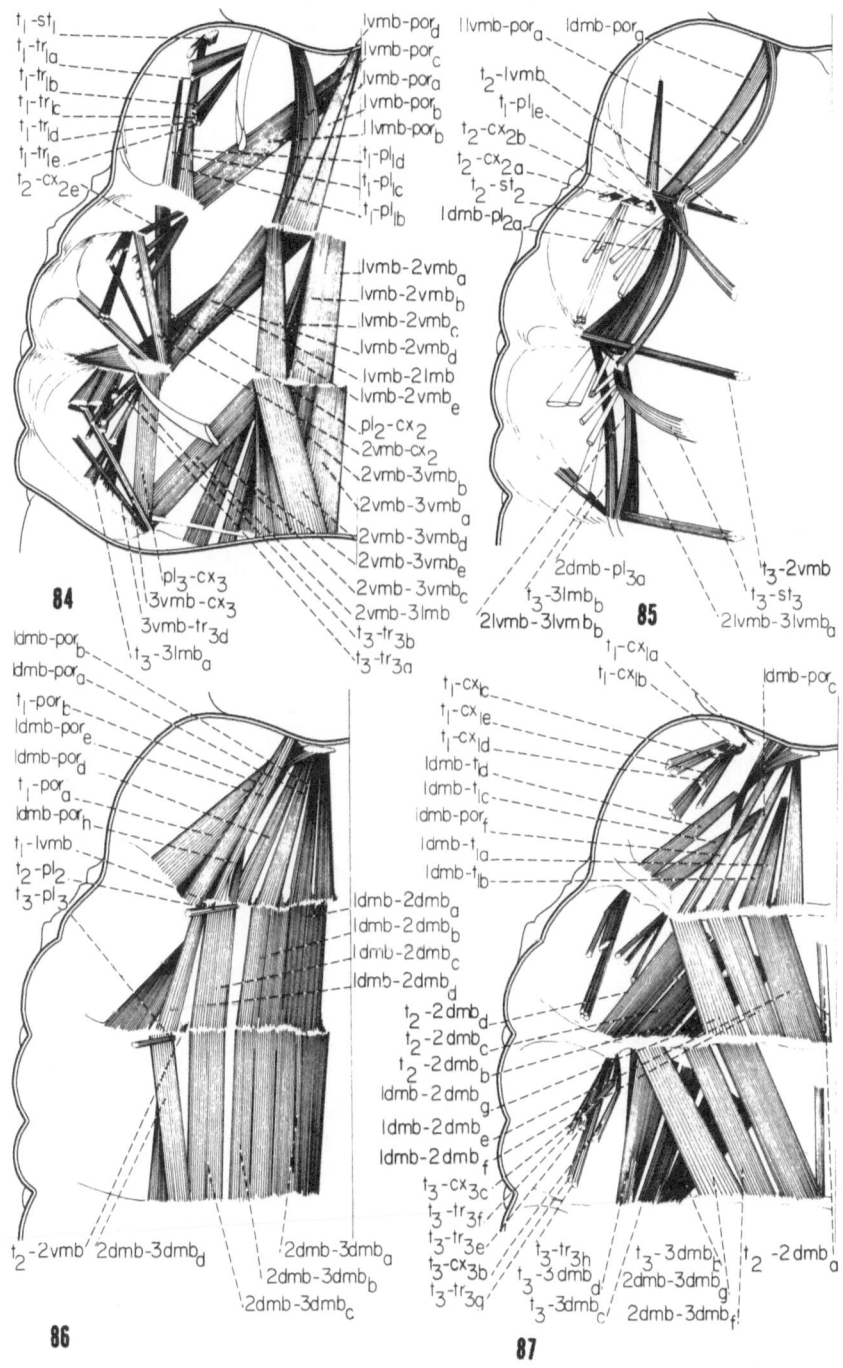

Figs. 84–87. Thorax of FG larva. Ventral view (right half) (series concluded).

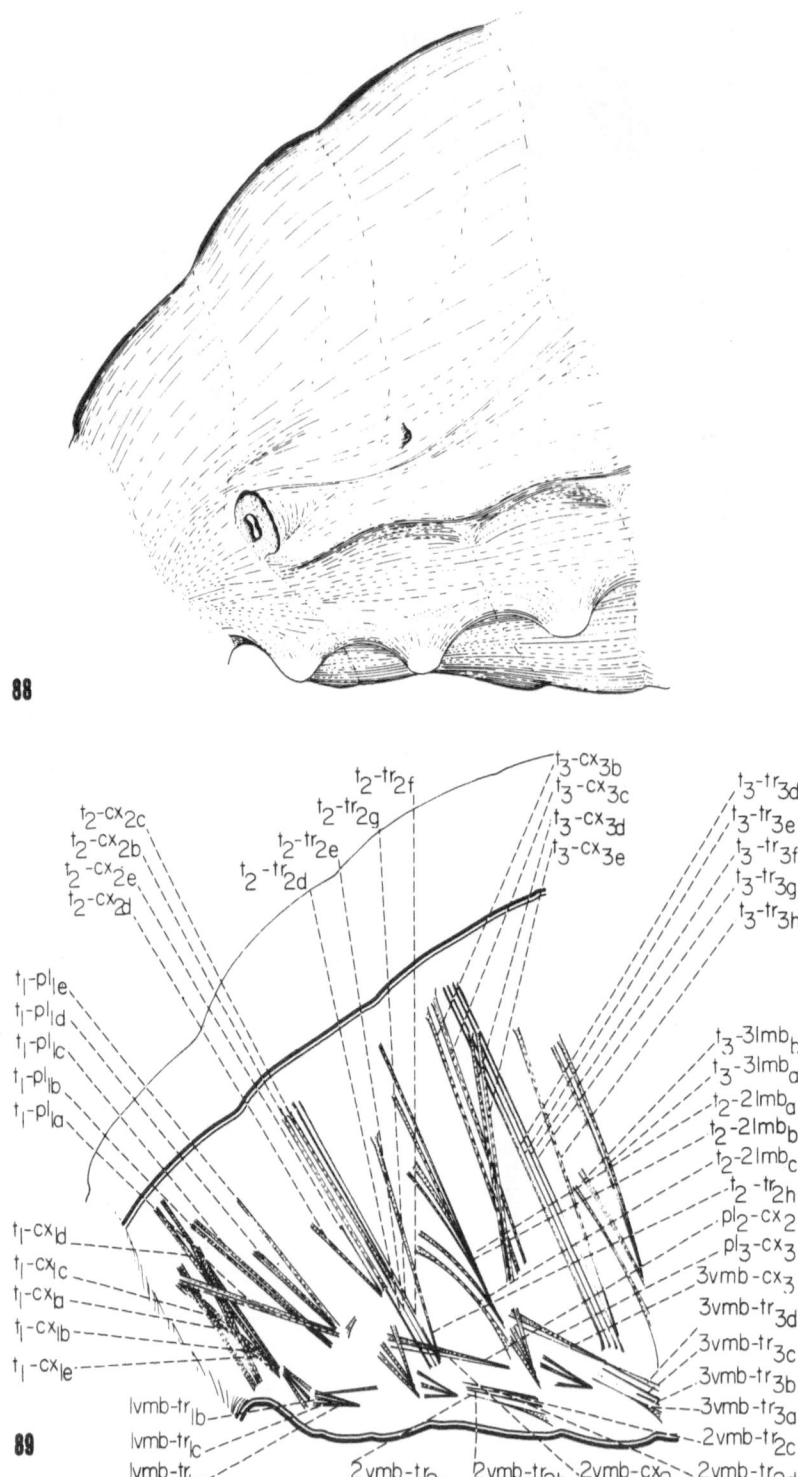

88

89

$t_2\text{-}cx_{2c}$
$t_2\text{-}cx_{2b}$
$t_2\text{-}cx_{2e}$
$t_2\text{-}cx_{2d}$

$t_2\text{-}tr_{2f}$
$t_2\text{-}tr_{2g}$
$t_2\text{-}tr_{2e}$
$t_2\text{-}tr_{2d}$

$t_3\text{-}cx_{3b}$
$t_3\text{-}cx_{3c}$
$t_3\text{-}cx_{3d}$
$t_3\text{-}cx_{3e}$

$t_3\text{-}tr_{3d}$
$t_3\text{-}tr_{3e}$
$t_3\text{-}tr_{3f}$
$t_3\text{-}tr_{3g}$
$t_3\text{-}tr_{3h}$

$t_1\text{-}pl_{1e}$
$t_1\text{-}pl_{1d}$
$t_1\text{-}pl_{1c}$
$t_1\text{-}pl_{1b}$
$t_1\text{-}pl_{1a}$

$t_3\text{-}3lmb_b$
$t_3\text{-}3lmb_a$
$t_2\text{-}2lmb_a$
$t_2\text{-}2lmb_b$
$t_2\text{-}2lmb_c$
$t_2\text{-}tr_{2h}$
$pl_2\text{-}cx_2$
$pl_3\text{-}cx_3$
$3vmb\text{-}cx_3$
$3vmb\text{-}tr_{3d}$
$3vmb\text{-}tr_{3c}$
$3vmb\text{-}tr_{3b}$
$3vmb\text{-}tr_{3a}$
$2vmb\text{-}tr_{2c}$

$t_1\text{-}cx_{1d}$
$t_1\text{-}cx_{1c}$
$t_1\text{-}cx_{1a}$
$t_1\text{-}cx_{1b}$
$t_1\text{-}cx_{1e}$

$lvmb\text{-}tr_{1b}$
$lvmb\text{-}tr_{1c}$
$lvmb\text{-}tr_{1a}$

$2vmb\text{-}tr_{2a}$
$2vmb\text{-}tr_{2b}$
$2vmb\text{-}cx_2$
$2vmb\text{-}tr_{2d}$

Figs. 88–89. Thorax of C larva. Lateral view (see also Figs. 90–91).

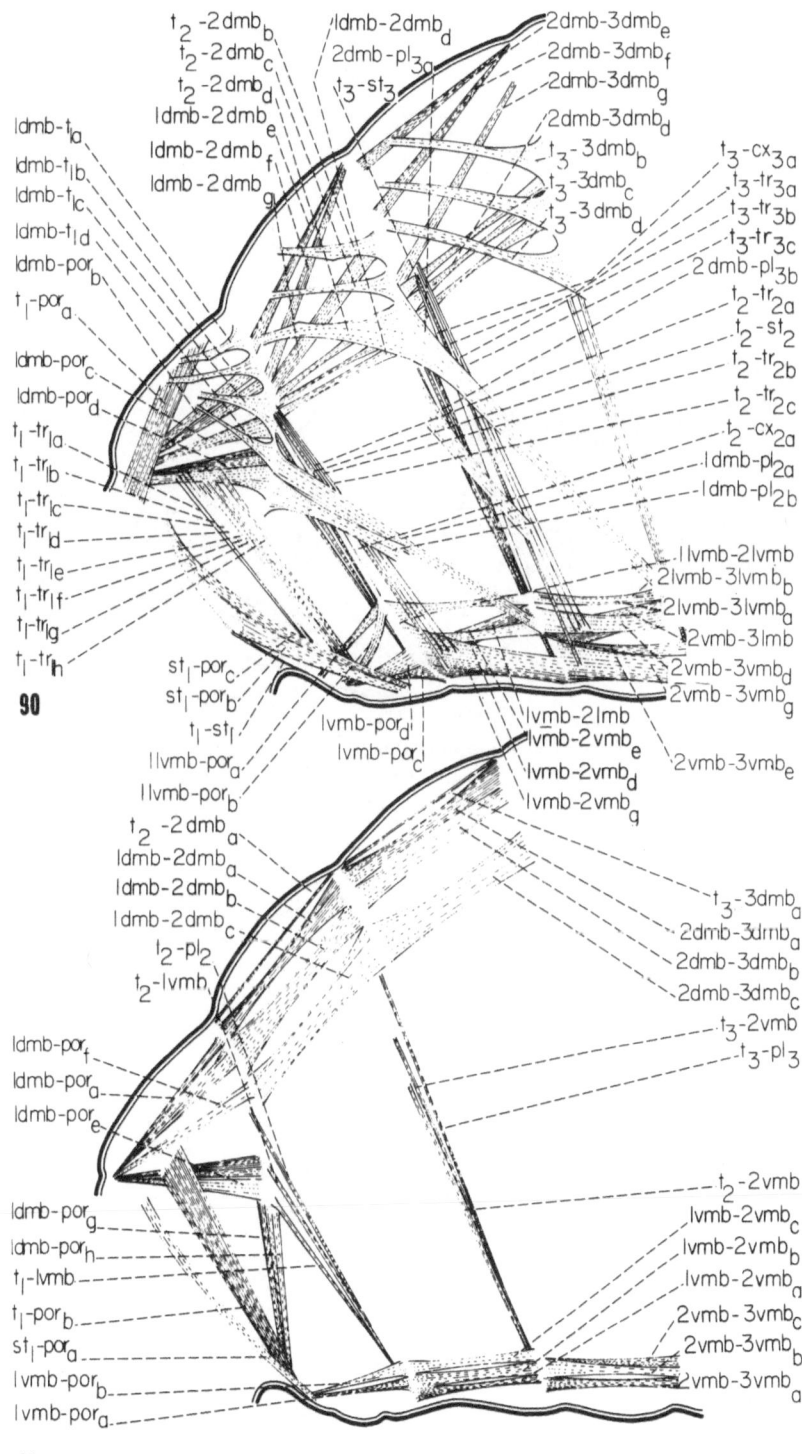

t_2-2dmb$_b$
t_2-2dmb$_c$
t_2-2dmb$_d$
ldmb-2dmb$_e$
ldmb-2dmb$_f$
ldmb-2dmb$_g$

ldmb-2dmb$_d$
2dmb-pl$_{3a}$
t_3-st$_3$

2dmb-3dmb$_e$
2dmb-3dmb$_f$
2dmb-3dmb$_g$
2dmb-3dmb$_d$
t_3-3dmb$_b$
t_3-3dmb$_c$
t_3-3dmb$_d$

ldmb-t$_{Ia}$
ldmb-t$_{Ib}$
ldmb-t$_{Ic}$
ldmb-t$_{Id}$
ldmb-por$_b$
t_1-por$_a$

ldmb-por$_c$
ldmb-por$_d$
t_1-tr$_{Ia}$
t_1-tr$_{Ib}$
t_1-tr$_{Ic}$
t_1-tr$_{Id}$
t_1-tr$_{Ie}$
t_1-tr$_{If}$
t_1-tr$_{Ig}$
t_1-tr$_{Ih}$

t_3-cx$_{3a}$
t_3-tr$_{3a}$
t_3-tr$_{3b}$
t_3-tr$_{3c}$
2dmb-pl$_{3b}$
t_2-tr$_{2a}$
t_2-st$_2$
t_2-tr$_{2b}$
t_2-tr$_{2c}$
t_2-cx$_{2a}$
ldmb-pl$_{2a}$
ldmb-pl$_{2b}$
llvmb-2lvmb
2lvmb-3lvmb$_b$
2lvmb-3lvmb$_a$
2vmb-3lmb
2vmb-3vmb$_d$
2vmb-3vmb$_g$

st$_1$-por$_c$
st$_1$-por$_b$
t_1-st$_1$
llvmb-por$_a$
llvmb-por$_b$

lvmb-por$_d$
lvmb-por$_c$

lvmb-2lmb
lvmb-2vmb$_e$
lvmb-2vmb$_d$
lvmb-2vmb$_g$

2vmb-3vmb$_e$

90

t_2-2dmb$_a$
ldmb-2dmb$_a$
ldmb-2dmb$_b$
ldmb-2dmb$_c$
t_2-pl$_2$
t_2-lvmb

t_3-3dmb$_a$
2dmb-3dmb$_a$
2dmb-3dmb$_b$
2dmb-3dmb$_c$
t_3-2vmb
t_3-pl$_3$

ldmb-por$_f$
ldmb-por$_a$
ldmb-por$_e$

ldmb-por$_g$
ldmb-por$_h$
t_1-lvmb
t_1-por$_b$
st$_1$-por$_a$
lvmb-por$_b$
lvmb-por$_a$

t_2-2vmb
lvmb-2vmb$_c$
lvmb-2vmb$_b$
lvmb-2vmb$_a$
2vmb-3vmb$_c$
2vmb-3vmb$_b$
2vmb-3vmb$_a$

91

Figs. 90–91. Thorax of C larva. Lateral view (series concluded).

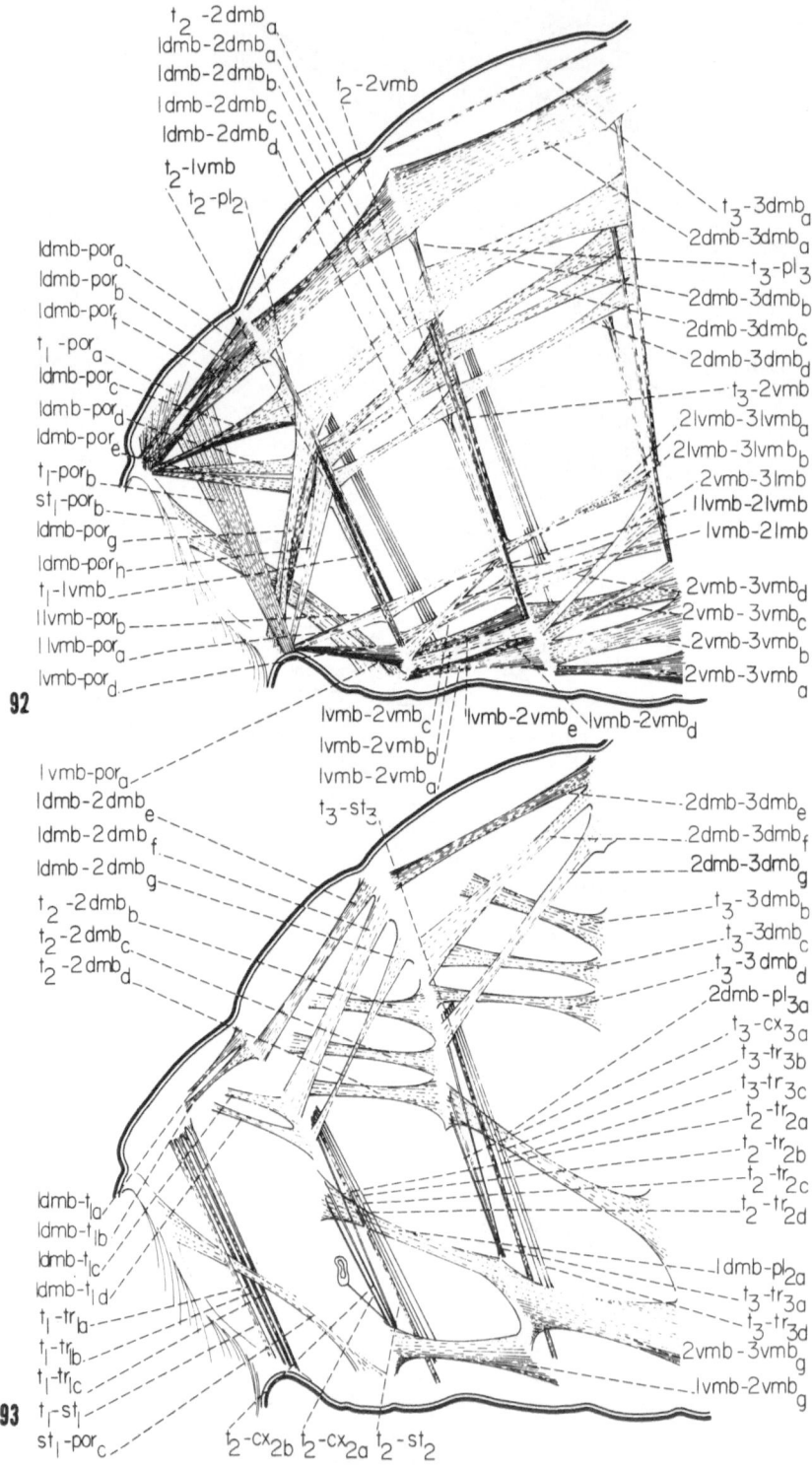

t_2-2dmb$_a$
ldmb-2dmb$_a$
ldmb-2dmb$_b$
ldmb-2dmb$_c$
ldmb-2dmb$_d$
t_2-lvmb
t_2-pl$_2$

t_2-2vmb

t_3-3dmb$_a$
2dmb-3dmb$_a$
t_3-pl$_3$
2dmb-3dmb$_b$
2dmb-3dmb$_c$
2dmb-3dmb$_d$
t_3-2vmb
2lvmb-3lvmb$_a$
2lvmb-3lvmb$_b$
2vmb-3lmb
1lvmb-2lvmb
lvmb-2lmb

ldmb-por$_a$
ldmb-por$_b$
ldmb-por$_f$
t_1-por$_a$
ldmb-por$_c$
ldmb-por$_d$
ldmb-por$_e$
t_1-por$_b$
st$_1$-por$_b$
ldmb-por$_g$
ldmb-por$_h$
t_1-lvmb
1lvmb-por$_b$
1lvmb-por$_a$
lvmb-por$_d$

2vmb-3vmb$_d$
2vmb-3vmb$_c$
2vmb-3vmb$_b$
2vmb-3vmb$_a$

92

lvmb-2vmb$_c$ lvmb-2vmb$_e$ lvmb-2vmb$_d$
lvmb-2vmb$_b$
lvmb-2vmb$_a$

1vmb-por$_a$
ldmb-2dmb$_e$
ldmb-2dmb$_f$
ldmb-2dmb$_g$
t_2-2dmb$_b$
t_2-2dmb$_c$
t_2-2dmb$_d$

t_3-st$_3$

2dmb-3dmb$_e$
2dmb-3dmb$_f$
2dmb-3dmb$_g$
t_3-3dmb$_b$
t_3-3dmb$_c$
t_3-3dmb$_d$
2dmb-pl$_{3a}$
t_3-cx$_{3a}$
t_3-tr$_{3b}$
t_3-tr$_{3c}$
t_2-tr$_{2a}$
t_2-tr$_{2b}$
t_2-tr$_{2c}$
t_2-tr$_{2d}$

ldmb-t$_{1a}$
ldmb-t$_{1b}$
ldmb-t$_{1c}$
ldmb-t$_{1d}$
t_1-tr$_{1a}$
t_1-tr$_{1b}$
t_1-tr$_{1c}$

ldmb-pl$_{2a}$
t_3-tr$_{3a}$
t_3-tr$_{3d}$
2vmb-3vmb$_g$
lvmb-2vmb$_g$

93 t_1-st$_1$
st$_1$-por$_c$

t_2-cx$_{2b}$ t_2-cx$_{2a}$ t_2-st$_2$

Figs. 92–93.   Thorax of C larva. Sagittal view (see also Figs. 94–95).

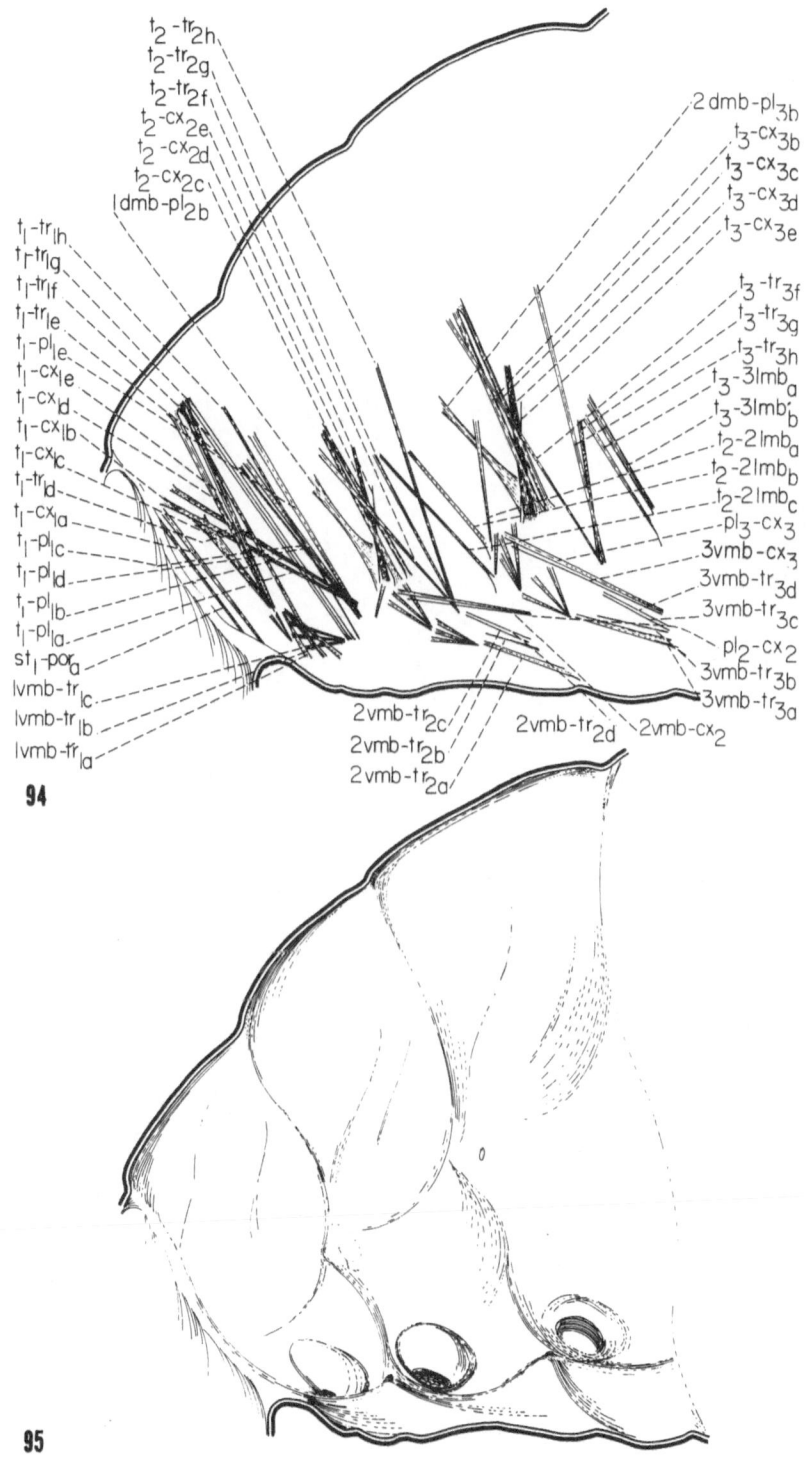

t_2-tr_{2h}
t_2-tr_{2g}
t_2-tr_{2f}
t_2-cx_{2e}
t_2-cx_{2d}
t_2-cx_{2c}
$1dmb$-pl_{2b}

t_1-tr_{1h}
t_1-tr_{1g}
t_1-tr_{1f}
t_1-tr_{1e}
t_1-pl_{1e}
t_1-cx_{1e}
t_1-cx_{1d}
t_1-cx_{1b}
t_1-cx_{1c}
t_1-tr_{1d}
t_1-cx_{1a}
t_1-pl_{1c}
t_1-pl_{1d}
t_1-pl_{1b}
t_1-pl_{1a}
st_1-po_{1a}
$lvmb$-tr_{1c}
$lvmb$-tr_{1b}
$lvmb$-tr_{1a}

$2dmb$-pl_{3b}
t_3-cx_{3b}
t_3-cx_{3c}
t_3-cx_{3d}
t_3-cx_{3e}

t_3-tr_{3f}
t_3-tr_{3g}
t_3-tr_{3h}
t_3-$3lmb_a$
t_3-$3lmb'_b$
t_2-$2lmb_a$
t_2-$2lmb_b$
t_2-$2lmb_c$
pl_3-cx_3
$3vmb$-cx_3
$3vmb$-tr_{3d}
$3vmb$-tr_{3c}
pl_2-cx_2
$3vmb$-tr_{3b}
$3vmb$-tr_{3a}

$2vmb$-tr_{2c}
$2vmb$-tr_{2b}
$2vmb$-tr_{2a}
$2vmb$-tr_{2d}
$2vmb$-cx_2

94

95

Figs. 94–95. Thorax of C larva. Sagittal view (series concluded). All muscles removed in Fig. 95.

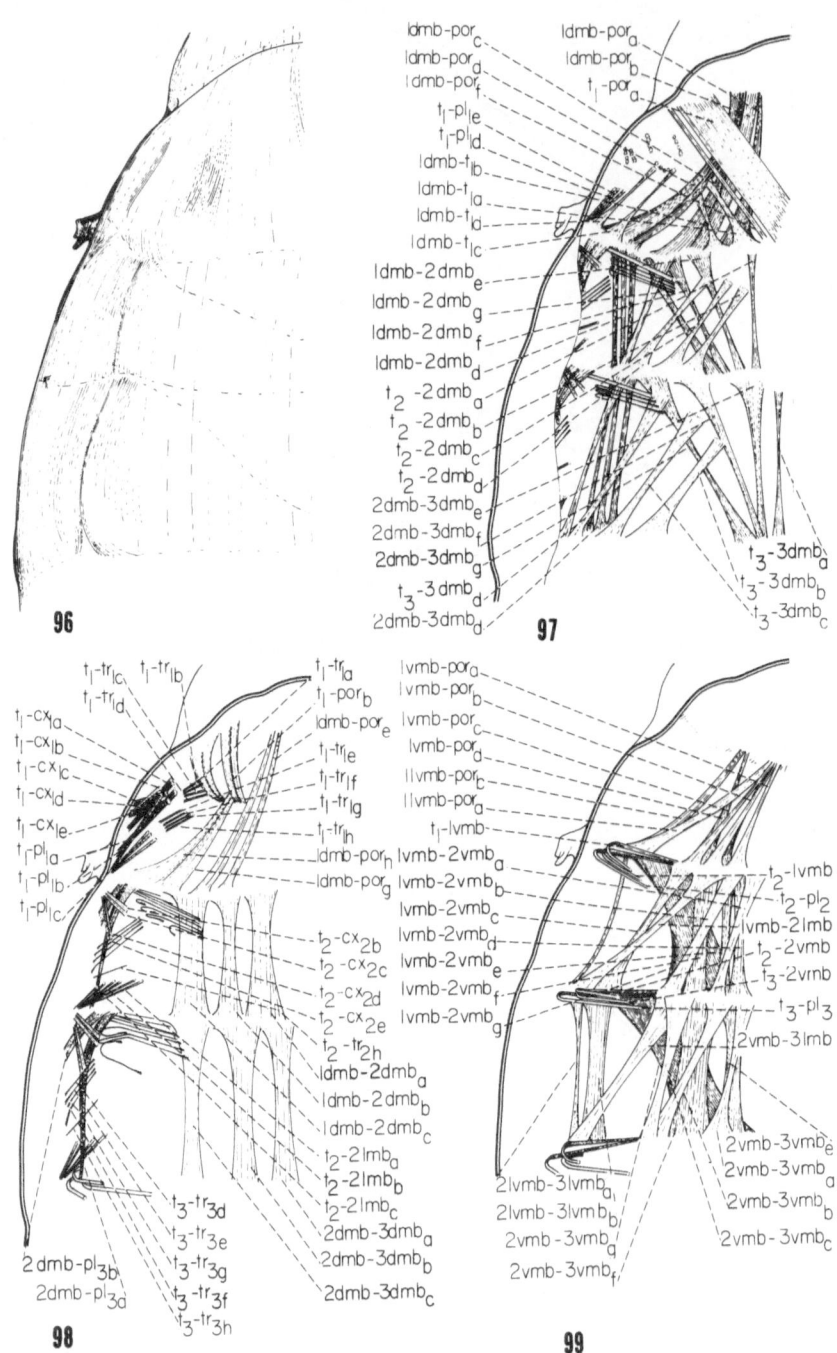

96

ldmb-por$_c$
ldmb-por$_d$
ldmb-por$_f$
t$_1$-pl$_e$
t$_1$-pl$_d$
ldmb-t$_{1b}$
ldmb-t$_{1a}$
ldmb-t$_{1d}$
ldmb-t$_{1c}$
ldmb-2dmb$_e$
ldmb-2dmb$_g$
ldmb-2dmb$_f$
ldmb-2dmb$_d$
t$_2$-2dmb$_a$
t$_2$-2dmb$_b$
t$_2$-2dmb$_c$
t$_2$-2dmb$_d$
2dmb-3dmb$_e$
2dmb-3dmb$_f$
2dmb-3dmb$_g$
t$_3$-3dmb$_d$
2dmb-3dmb$_d$

ldmb-por$_a$
ldmb-por$_b$
t$_1$-por$_a$

t$_3$-3dmb$_a$
t$_3$-3dmb$_b$
t$_3$-3dmb$_c$

97

t$_1$-tr$_{1c}$ t$_1$-tr$_{1b}$
t$_1$-tr$_{1d}$
t$_1$-cx$_{1a}$
t$_1$-cx$_{1b}$
t$_1$-cx$_{1c}$
t$_1$-cx$_{1d}$
t$_1$-cx$_{1e}$
t$_1$-pl$_a$
t$_1$-pl$_b$
t$_1$-pl$_c$

t$_2$-cx$_{2b}$
t$_2$-cx$_{2c}$
t$_2$-cx$_{2d}$
t$_2$-cx$_{2e}$
t$_2$-tr$_{2h}$
ldmb-2dmb$_a$
ldmb-2dmb$_b$
ldmb-2dmb$_c$
t$_2$-2lmb$_a$
t$_2$-2lmb$_b$
t$_2$-2lmb$_c$
2dmb-3dmb$_a$
2dmb-3dmb$_b$
2dmb-3dmb$_c$

t$_1$-tr$_{1a}$
t$_1$-por$_b$
ldmb-por$_e$
t$_1$-tr$_{1e}$
t$_1$-tr$_{1f}$
t$_1$-tr$_{1g}$
t$_1$-tr$_{1h}$
ldmb-por$_h$
ldmb-por$_g$

2dmb-pl$_{3b}$
2dmb-pl$_{3d}$
t$_3$-tr$_{3d}$
t$_3$-tr$_{3e}$
t$_3$-tr$_{3g}$
t$_3$-tr$_{3f}$
t$_3$-tr$_{3h}$

98

lvmb-por$_a$
lvmb-por$_b$
lvmb-por$_c$
lvmb-por$_d$
llvmb-por$_b$
llvmb-por$_a$
t$_1$-lvmb
lvmb-2vmb$_a$
lvmb-2vmb$_b$
lvmb-2vmb$_c$
lvmb-2vmb$_d$
lvmb-2vmb$_e$
lvmb-2vmb$_f$
lvmb-2vmb$_g$

2lvmb-3lvmb$_a$
2lvmb-3lvmb$_b$
2vmb-3vmb$_a$
2vmb-3vmb$_f$

t$_2$-lvmb
t$_2$-pl$_2$
lvmb-2lmb
t$_2$-2vmb
t$_3$-2vmb
t$_3$-pl$_3$
2vmb-3lmb

2vmb-3vmb$_e$
2vmb-3vmb$_a$
2vmb-3vmb$_b$
2vmb-3vmb$_c$

99

Figs. 96–99. Thorax of C larva. Dorsal view (left half) (see also Figs. 100–101).

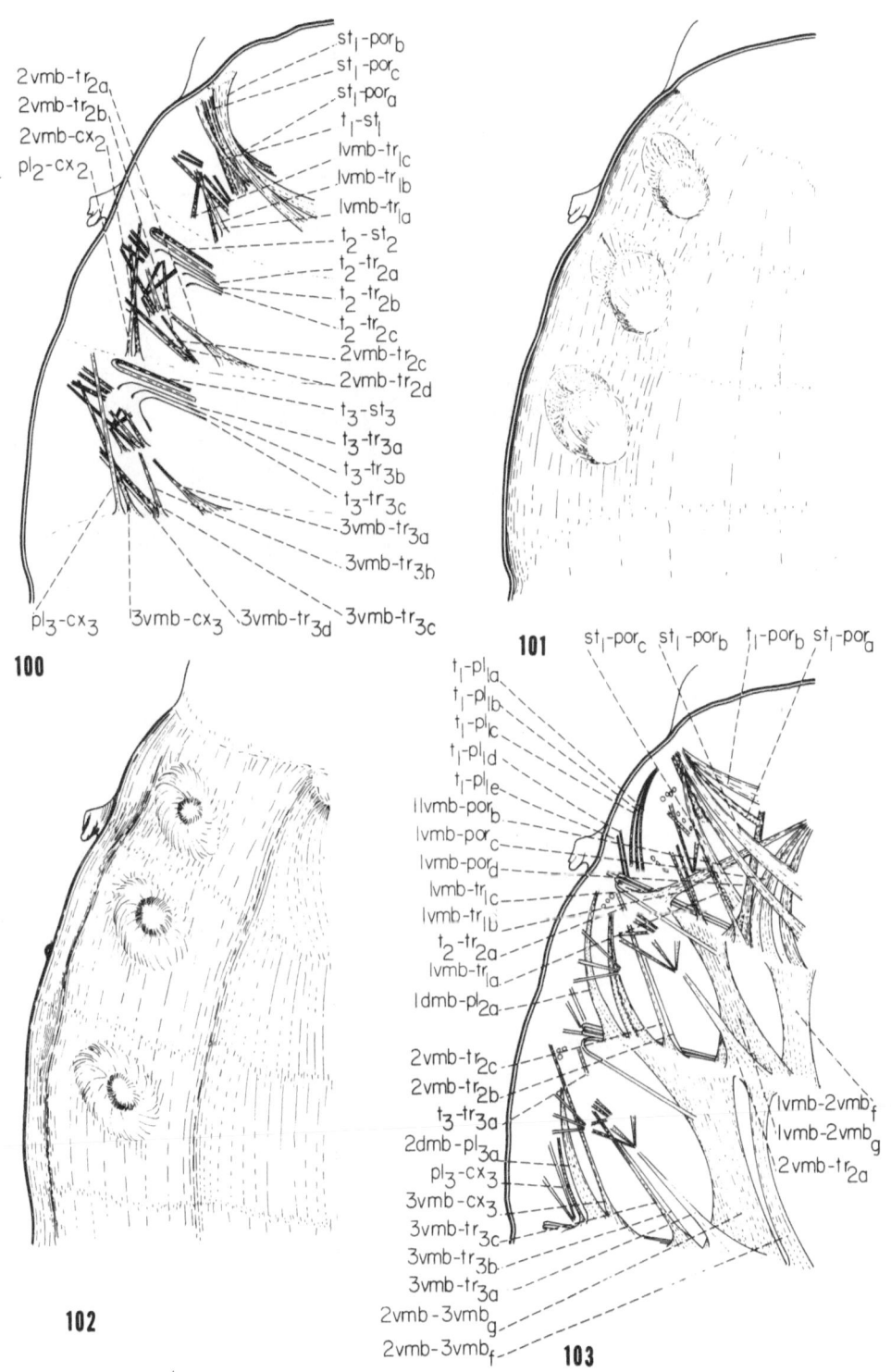

Figs. 100–103. Thorax of C larva. *Figs. 100–101.* Dorsal view (left half) (series concluded). *Figs. 102–103.* Ventral view (right half) (see also Figs. 104–106).

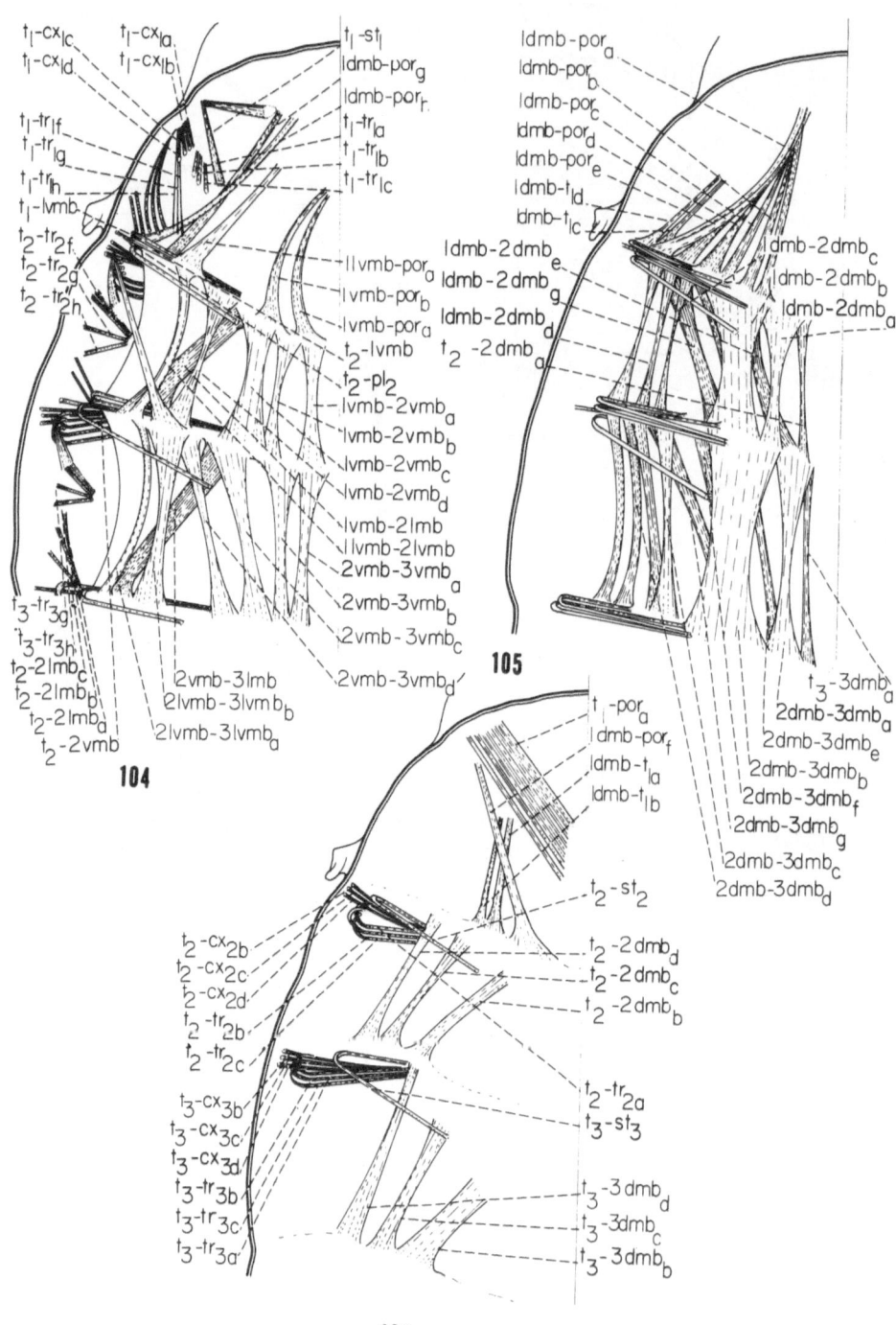

Figs. 104–106. Thorax of C larva. Ventral view (right half) (series concluded).

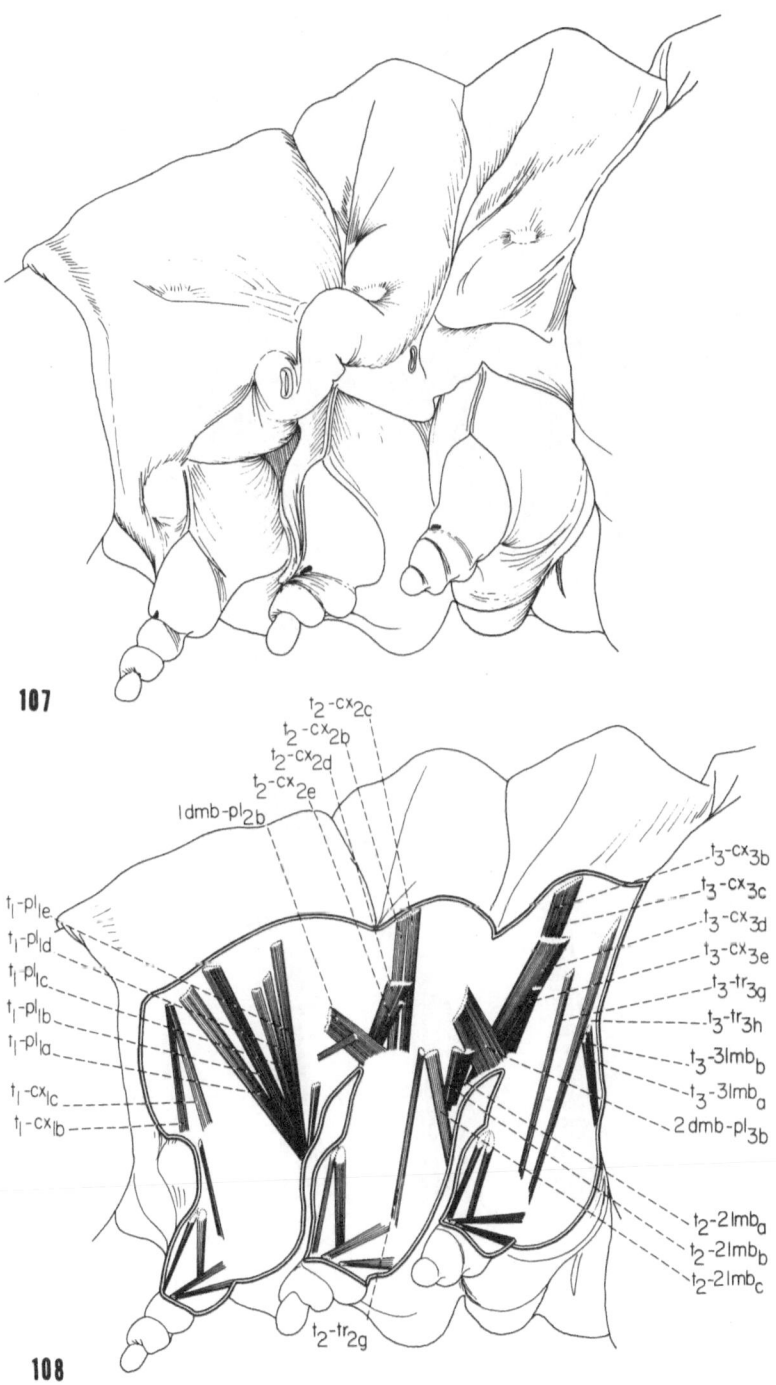

107

$t_2 - cx_{2c}$
$t_2 - cx_{2b}$
$t_2 - cx_{2d}$
$t_2 - cx_{2e}$
$l\,dmb - pl_{2b}$

$t_1 - pl_{1e}$
$t_1 - pl_{1d}$
$t_1 - pl_{1c}$
$t_1 - pl_{1b}$
$t_1 - pl_{1a}$

$t_1 - cx_{1c}$
$t_1 - cx_{1b}$

$t_3 - cx_{3b}$
$t_3 - cx_{3c}$
$t_3 - cx_{3d}$
$t_3 - cx_{3e}$
$t_3 - tr_{3g}$
$t_3 - tr_{3h}$
$t_3 - 3lmb_b$
$t_3 - 3lmb_a$
$2\,dmb - pl_{3b}$

$t_2 - 2lmb_a$
$t_2 - 2lmb_b$
$t_2 - 2lmb_c$

$t_2 - tr_{2g}$

108

Figs. 107–108. Thorax of C larva. Lateral view (see also Fig

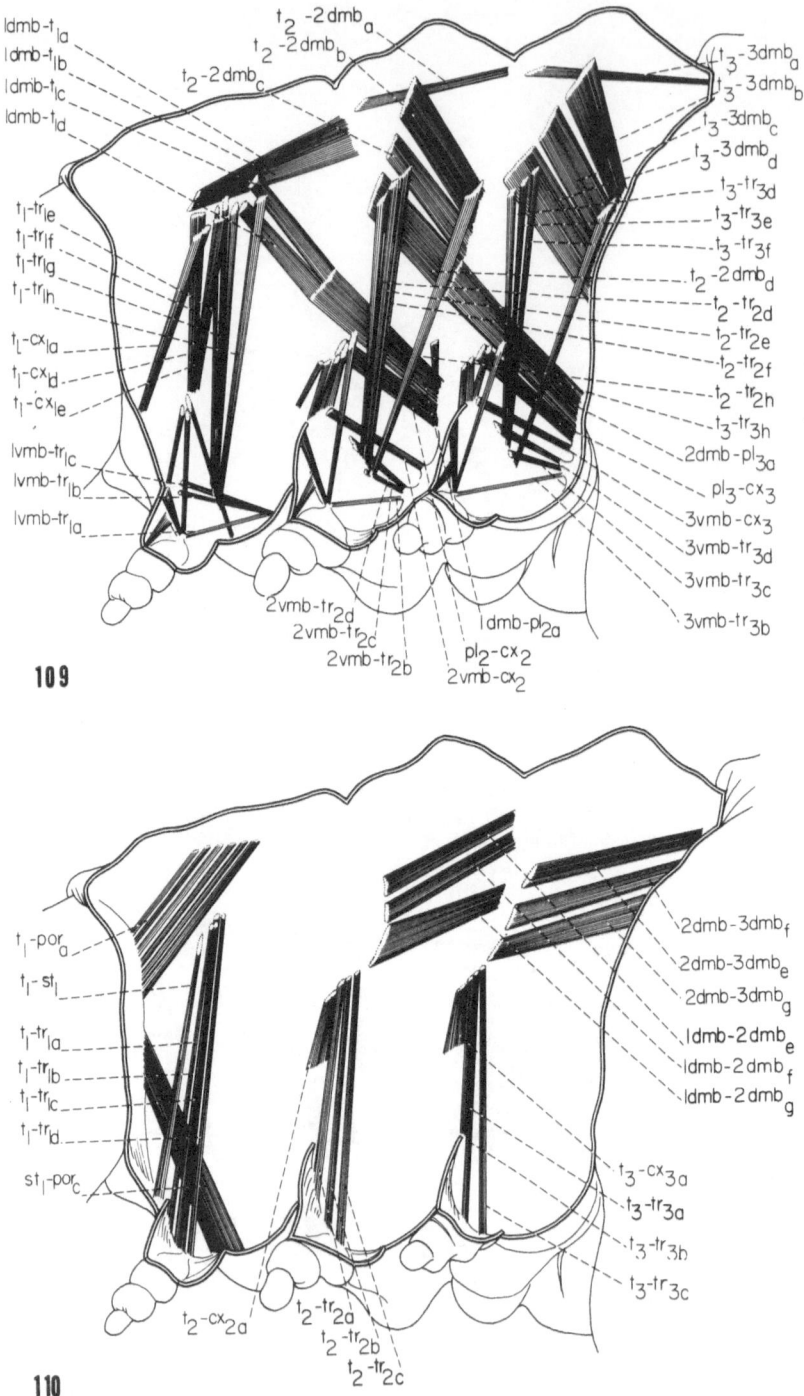

Figs. 109–110. Thorax of SG larva. Lateral view (series continued).

ldmb-por$_f$
ldmb-por$_a$
ldmb-por$_b$
ldmb-por$_c$

2dmb-3dmb$_a$
2dmb-3dmb$_b$
2dmb-3dmb$_c$
2dmb-3dmb$_d$

t$_1$-por$_b$
ldmb-por$_d$
ldmb-por$_e$

ldmb-2dmb$_a$
ldmb-2dmb$_b$
ldmb-2dmb$_c$
ldmb-2dmb$_d$

st$_1$-por$_a$
st$_1$-por$_b$

t$_3$-st$_3$
2lvmb-3lvmb$_b$
2vmb-3vmb$_g$
2vmb-3vmb$_f$
2vmb-3vmb$_e$
3vmb-tr$_{3a}$

llvmb-por$_b$
lvmb-por$_d$
lvmb-por$_c$

111

t$_2$-st$_2$
lvmb-2vmb$_f$

lvmb-2vmb$_e$
2vmb-tr$_{2a}$

lvmb-2vmb$_g$

t$_2$-2vmb
t$_2$-pl$_2$
t$_2$-lvmb

t$_3$-pl$_3$
t$_3$-2vmb

ldmb-por$_g$
ldmb-por$_h$
t$_1$-lvmb
llvmb-por$_a$
lvmb-por$_b$
lvmb-por$_a$

2lvmb-3lvmb$_a$
2vmb-3vmb$_d$
2vmb-3lmb
2vmb-3vmb$_c$
2vmb-3vmb$_b$
2vmb-3vmb$_a$

112

llvmb-2lvmb
lvmb-2vmb$_b$
lvmb-2vmb$_a$

lvmb-2vmb$_d$

lvmb-2lmb
lvmb-2vmb$_c$

Figs. 111–112. **Thorax of SG larva. Lateral view (series concluded).**

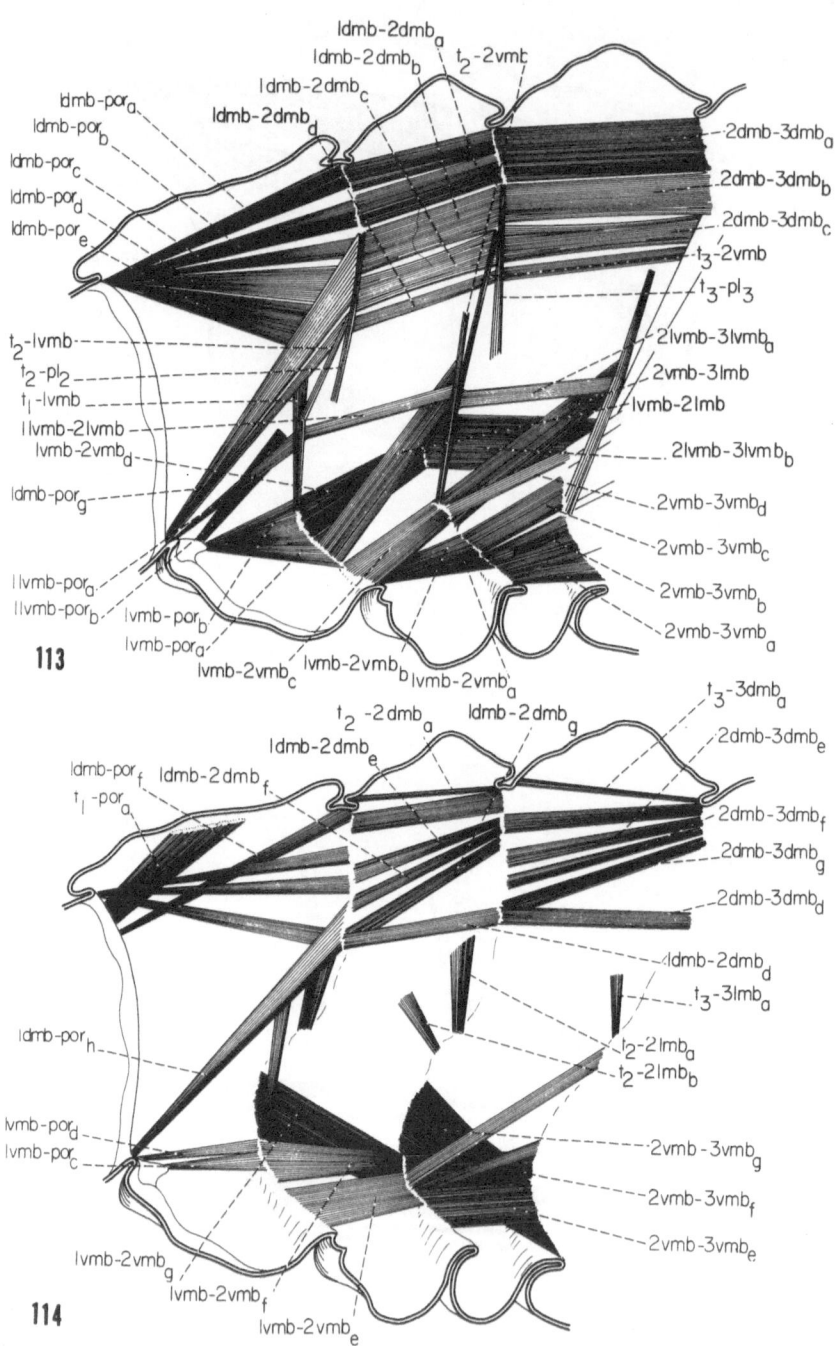

Figs. 113–114. Thorax of SG larva. Sagittal view (see also Figs. 115–118).

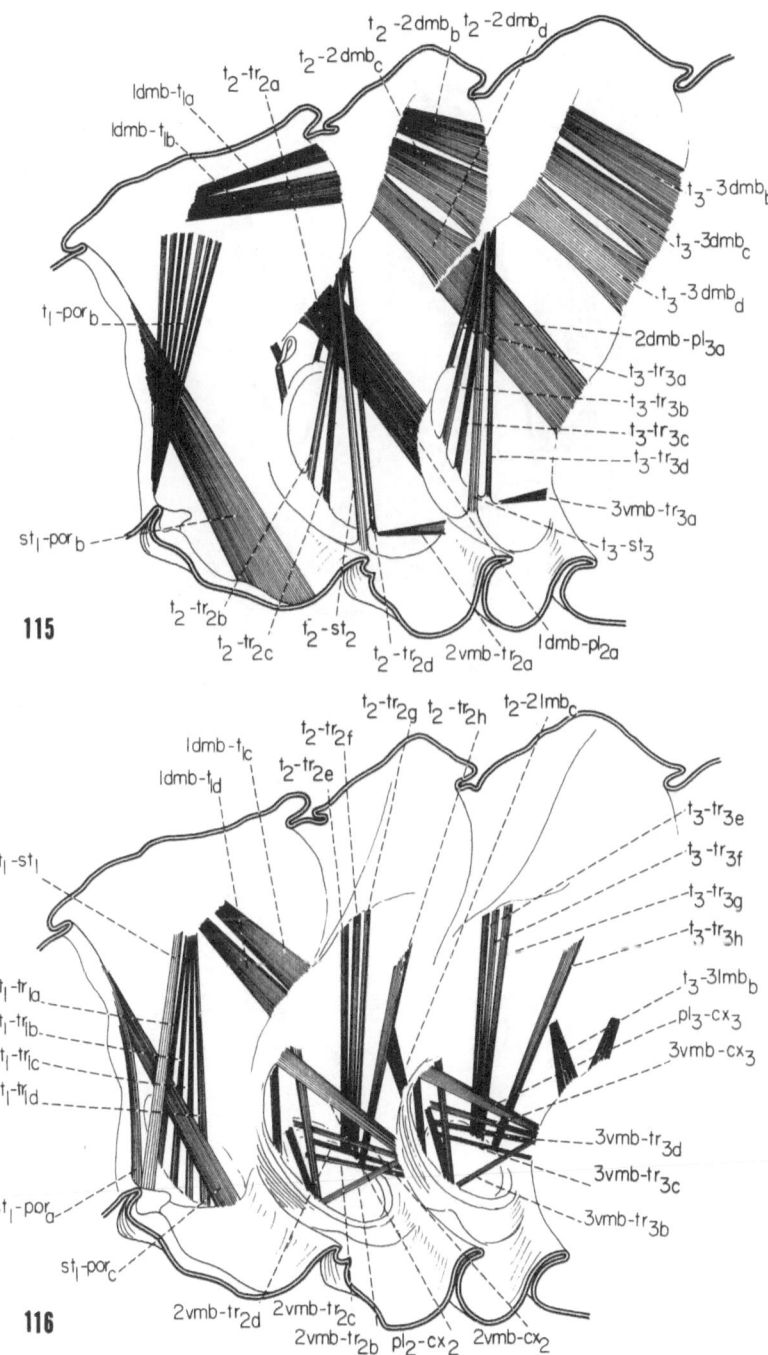

115

116

Figs. 115–116. **Thorax of SG larva. Sagittal view (series**

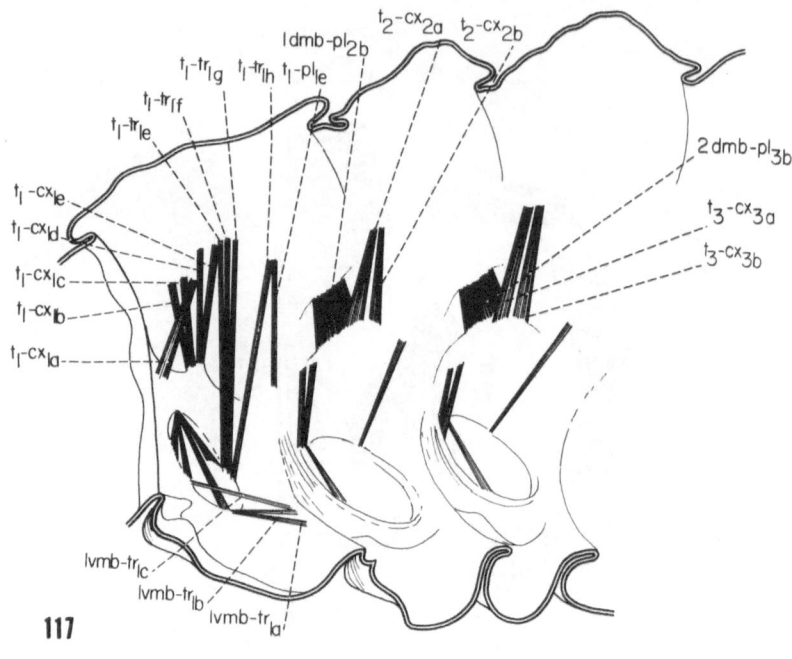

t_1-tr_{1g} t_1-tr_{1h} t_1-pl_{1e}

t_1-tr_{1f}

t_1-tr_{1e}

t_1-cx_{1e}

t_1-cx_{1d}

t_1-cx_{1c}

t_1-cx_{1b}

t_1-cx_{1a}

1dmb-pl_{2b}

t_2-cx_{2a} t_2-cx_{2b}

2dmb-pl_{3b}

t_3-cx_{3a}

t_3-cx_{3b}

1vmb-tr_{1c}

1vmb-tr_{1b} 1vmb-tr_{1a}

117

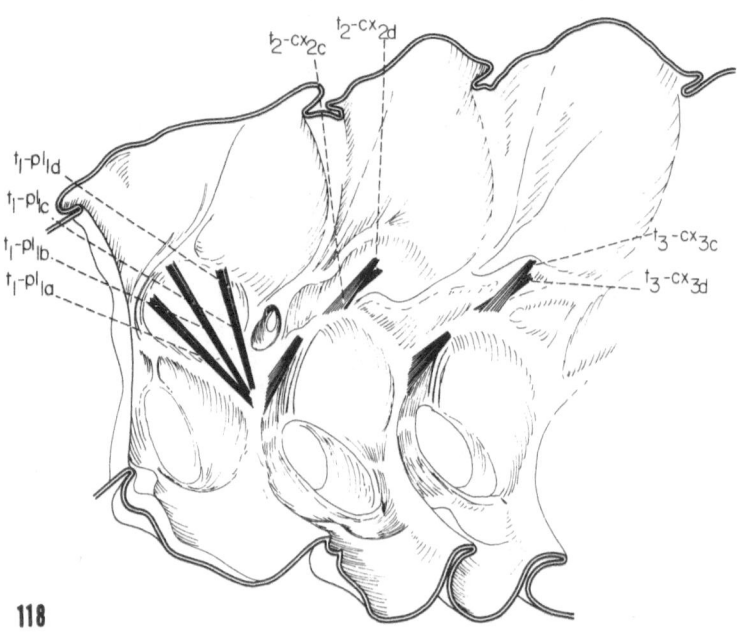

t_2-cx_{2c} t_2-cx_{2d}

t_1-pl_{1d}

t_1-pl_{1c}

t_1-pl_{1b}

t_1-pl_{1a}

t_3-cx_{3c}

t_3-cx_{3d}

118

Figs. 117–118. Thorax of SG larva. Sagittal view (series concluded).

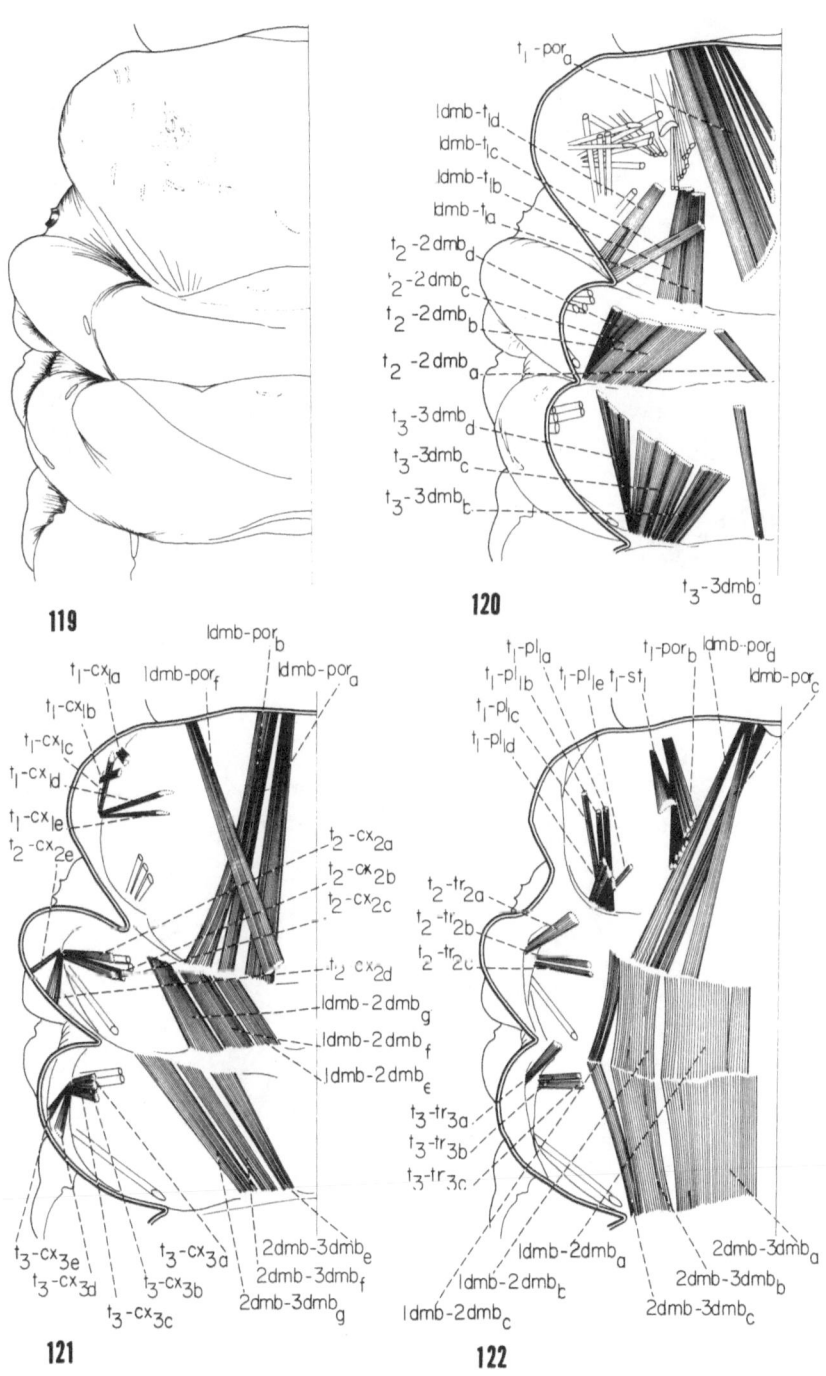

119

120

t_1-por_a
$ldmb$-t_{ld}
$ldmb$-t_{lc}
$ldmb$-t_{lb}
$ldmb$-t_{la}
t_2-$2dmb_d$
t_2-$2dmb_c$
t_2-$2dmb_b$
t_2-$2dmb_a$
t_3-$3dmb_d$
t_3-$3dmb_c$
t_3-$3dmb_b$
t_3-$3dmb_a$

121

$ldmb$-por_b
t_1-cx_{la} $ldmb$-por_f $ldmb$-por_a
t_1-cx_{lb}
t_1-cx_{lc}
t_1-cx_{ld}
t_1-cx_{le}
t_2-cx_{2e}
t_2-cx_{2a}
t_2-cx_{2b}
t_2-cx_{2c}
t_2-cx_{2d}
$ldmb$-$2dmb_g$
$ldmb$-$2dmb_f$
$ldmb$-$2dmb_e$
t_3-cx_{3e} t_3-cx_{3a} $2dmb$-$3dmb_e$
t_3-cx_{3d} t_3-cx_{3b} $2dmb$-$3dmb_f$
t_3-cx_{3c} $2dmb$-$3dmb_g$

122

t_1-pl_{la} t_1-por_b $ldmb$-por_d
t_1-pl_{lb} t_1-pl_{le} t_1-st_1 $ldmb$-por_c
t_1-pl_{lc}
t_1-pl_{ld}
t_2-tr_{2a}
t_2-tr_{2b}
t_2-tr_{2c}
t_3-tr_{3a}
t_3-tr_{3b}
t_3-tr_{3c}
$ldmb$-$2dmb_a$ $2dmb$-$3dmb_a$
$ldmb$-$2dmb_b$ $2dmb$-$3dmb_b$
$ldmb$-$2dmb_c$ $2dmb$-$3dmb_c$

Figs. 119–122. Thorax of SG larva. Dorsal view (left half) (see also Figs. 123–126).

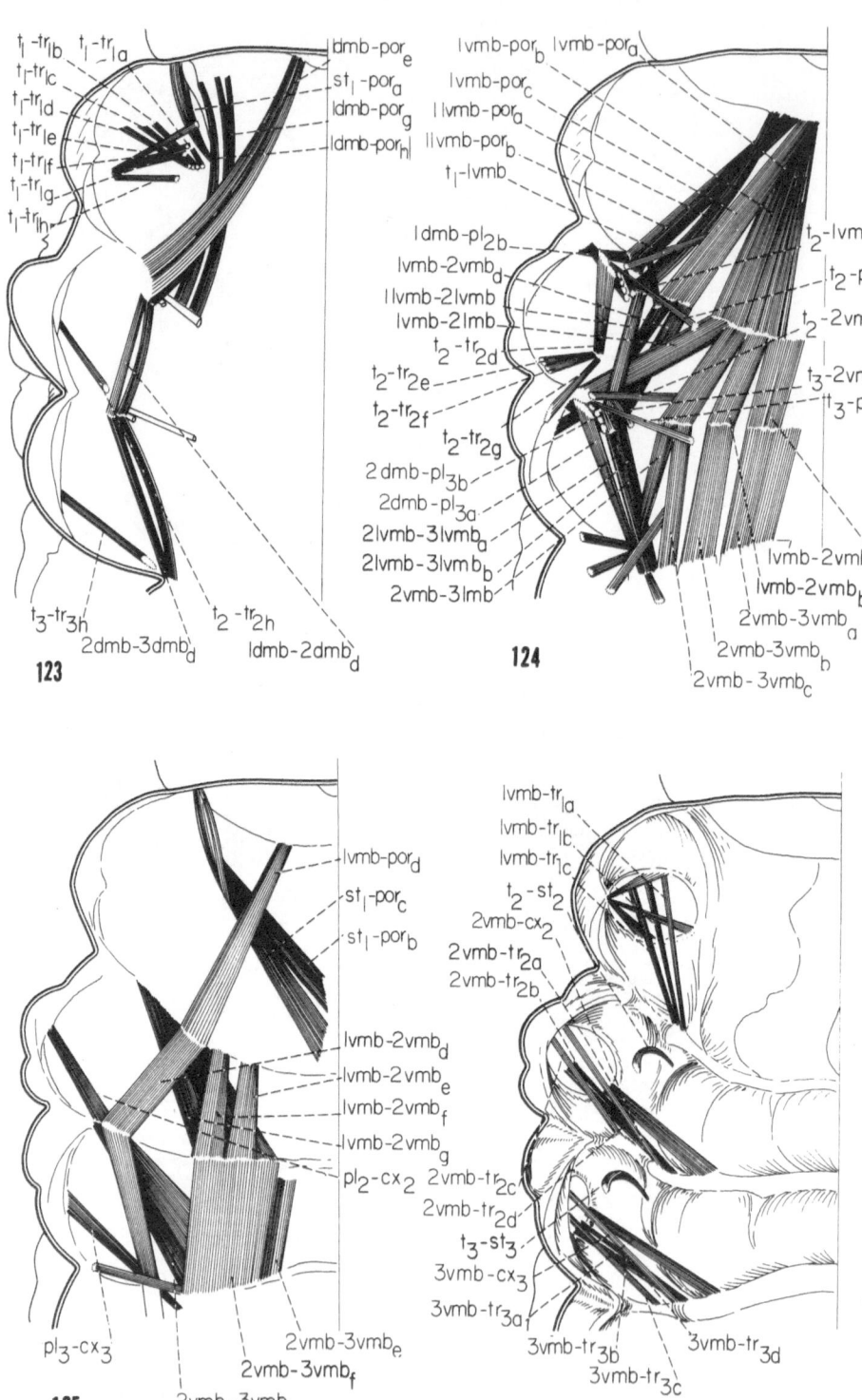

Figs. 123–126. Thorax of SG larva. Dorsal view (left half) (series concluded).

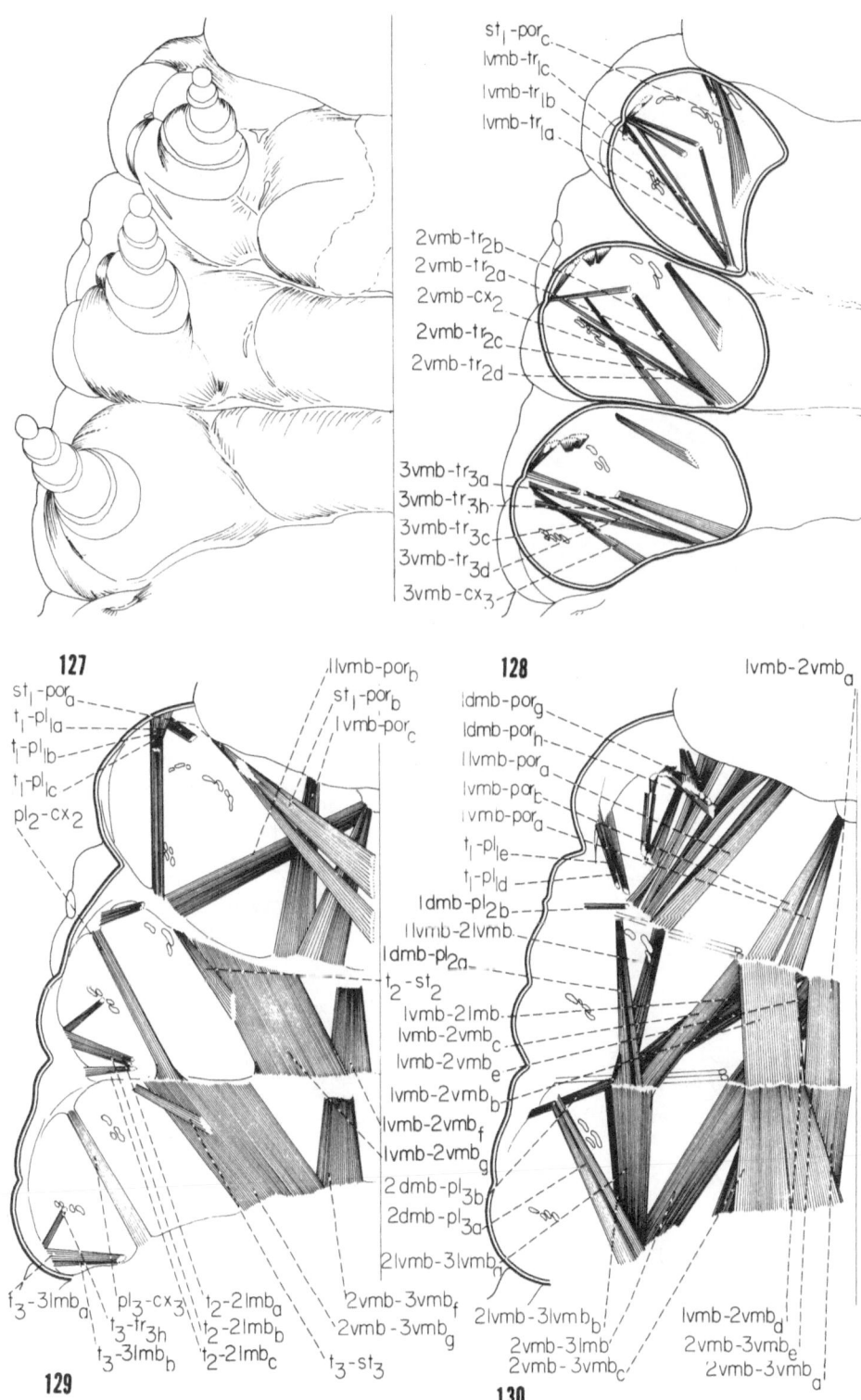

127

128

st$_1$-por$_c$
lvmb-tr$_{1c}$
lvmb-tr$_{1b}$
lvmb-tr$_{1a}$

2vmb-tr$_{2b}$
2vmb-tr$_{2a}$
2vmb-cx$_2$
2vmb-tr$_{2c}$
2vmb-tr$_{2d}$

3vmb-tr$_{3a}$
3vmb-tr$_{3h}$
3vmb-tr$_{3c}$
3vmb-tr$_{3d}$
3vmb-cx$_3$

st$_1$-por$_a$
t$_1$-pl$_{1a}$
t$_1$-pl$_{1b}$
t$_1$-pl$_{1c}$
pl$_2$-cx$_2$

1lvmb-por$_b$
st$_1$-por$_b$
lvmb-por$_c$

lvmb-2vmb$_a$

ldmb-por$_g$
ldmb-por$_h$
1lvmb-por$_a$
lvmb-por$_t$
lvmb-por$_a$
t$_1$-pl$_{1e}$
t$_1$-pl$_{1d}$
ldmb-pl$_{2b}$
1lvmb-2lvmb
ldmb-pl$_{2a}$
t$_2$-st$_2$
lvmb-2lmb
lvmb-2vmb$_c$
lvmb-2vmb$_e$
lvmb-2vmb$_b$
lvmb-2vmb$_f$
lvmb-2vmb$_g$
2dmb-pl$_{3b}$
2dmb-pl$_{3a}$
2lvmb-3lvmb$_a$
2vmb-3vmb$_f$
2vmb-3vmb$_g$

129

t$_3$-3lmb$_a$
t$_3$-tr$_{3h}$
t$_3$-3lmb$_b$
pl$_3$-cx$_3$
t$_2$-2lmb$_a$
t$_2$-2lmb$_b$
t$_2$-2lmb$_c$
t$_3$-st$_3$

130

2lvmb-3lvmb$_b$
2vmb-3lmb
2vmb-3vmb$_c$

lvmb-2vmb$_d$
2vmb-3vmb$_e$
2vmb-3vmb$_a$

Figs. 127–130. Thorax of SG larva. Ventral view (right half) (see also Figs. 131–134).

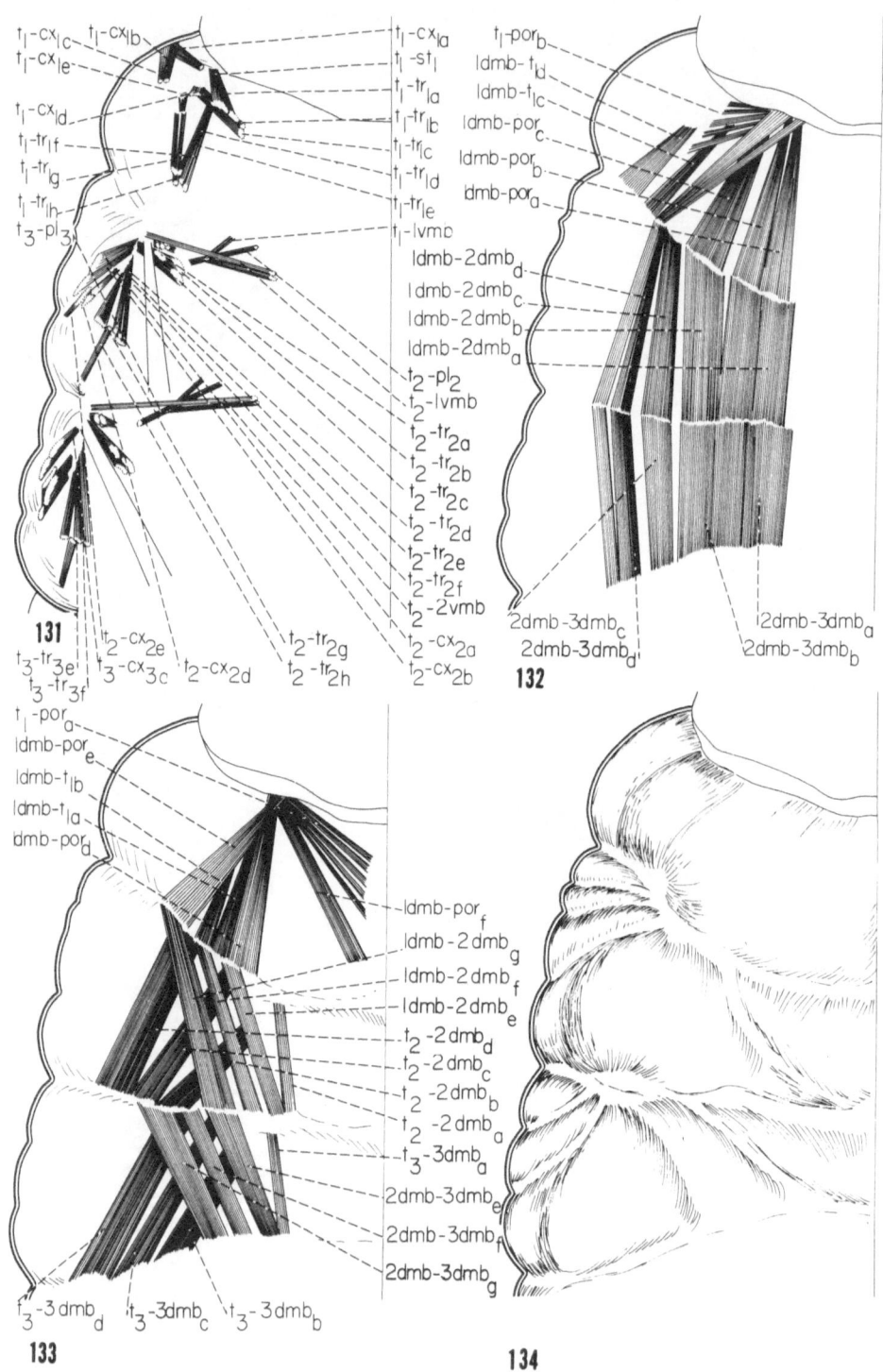

Figs. 131–134. Thorax of SG larva. Ventral view (right half) (series concluded). All muscles removed in Fig. 134.

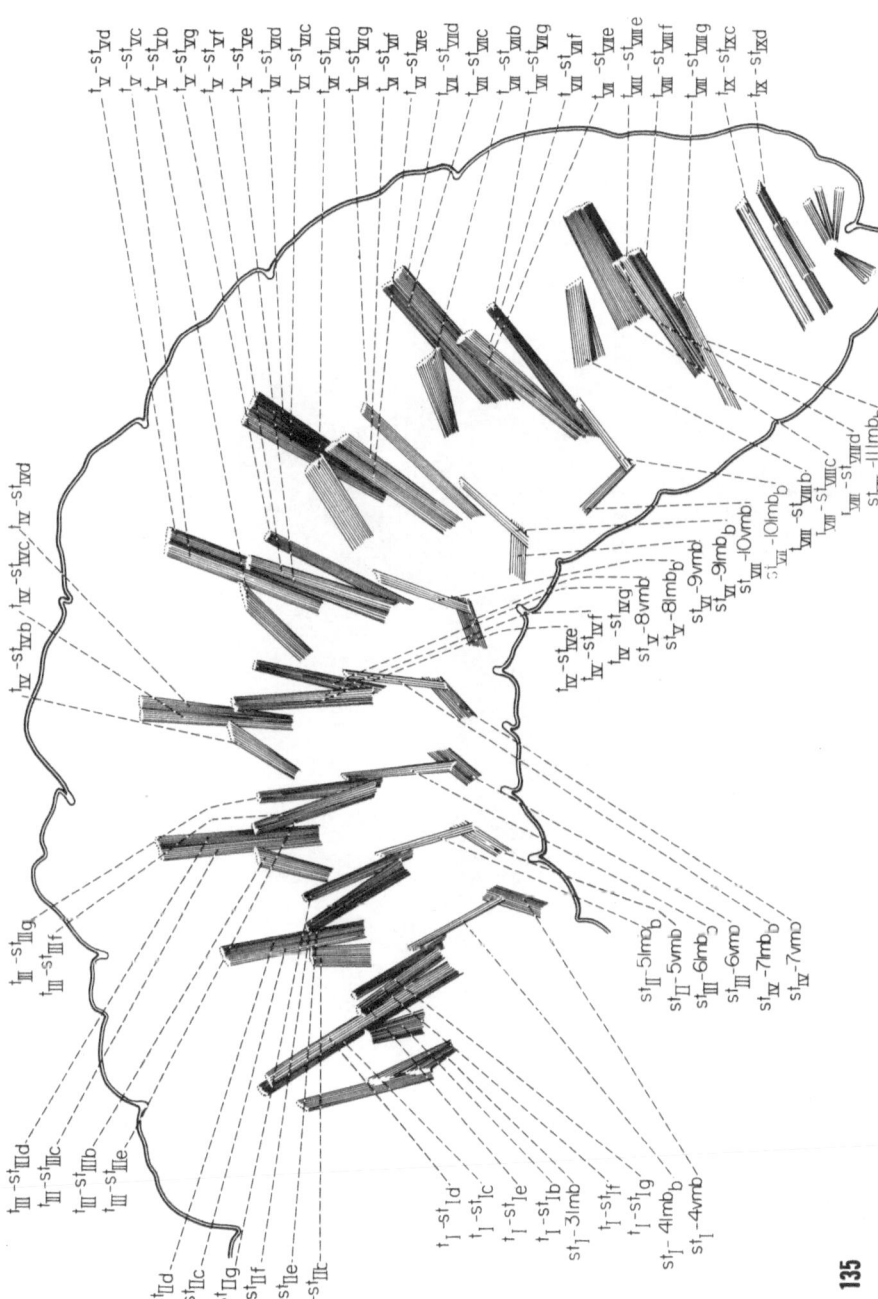

Fig. 135. Abdomen of FG larva. Lateral view (see also Figs. 136–137).

135

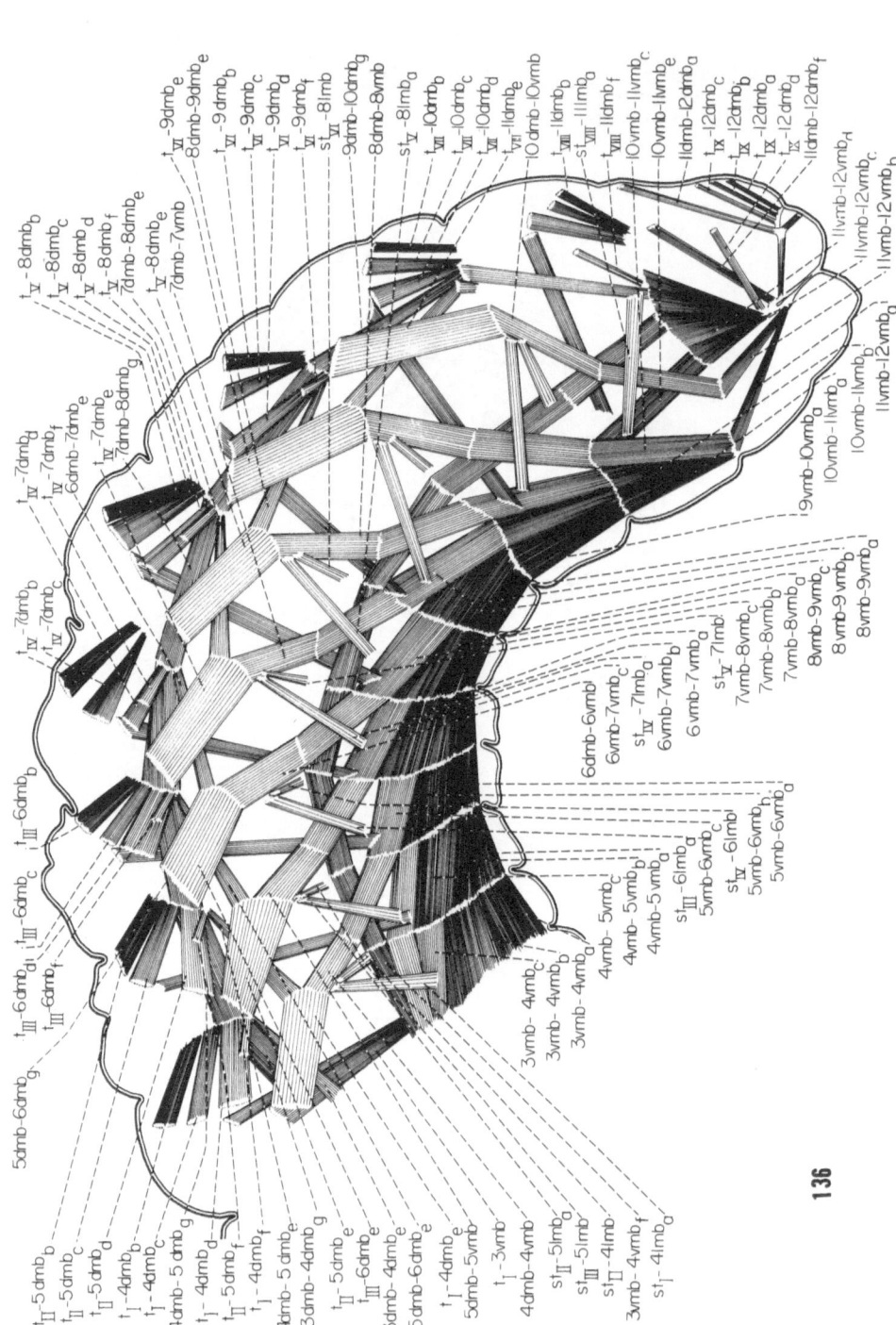

Fig. 136. Abdomen of FG larva. Lateral view (series continued).

136

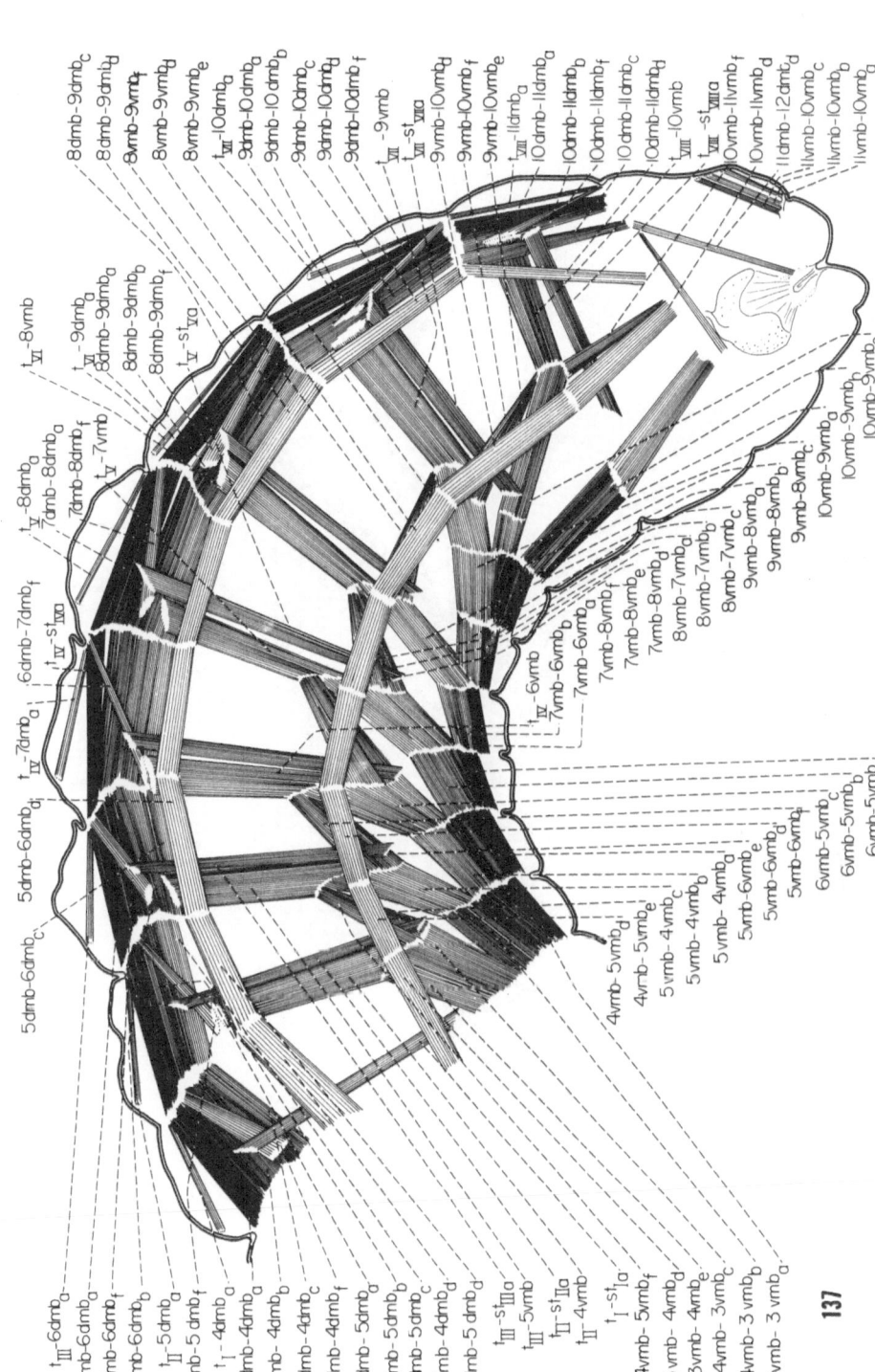

Fig. 137. Abdomen of FG larva. Lateral view (series concluded).

137

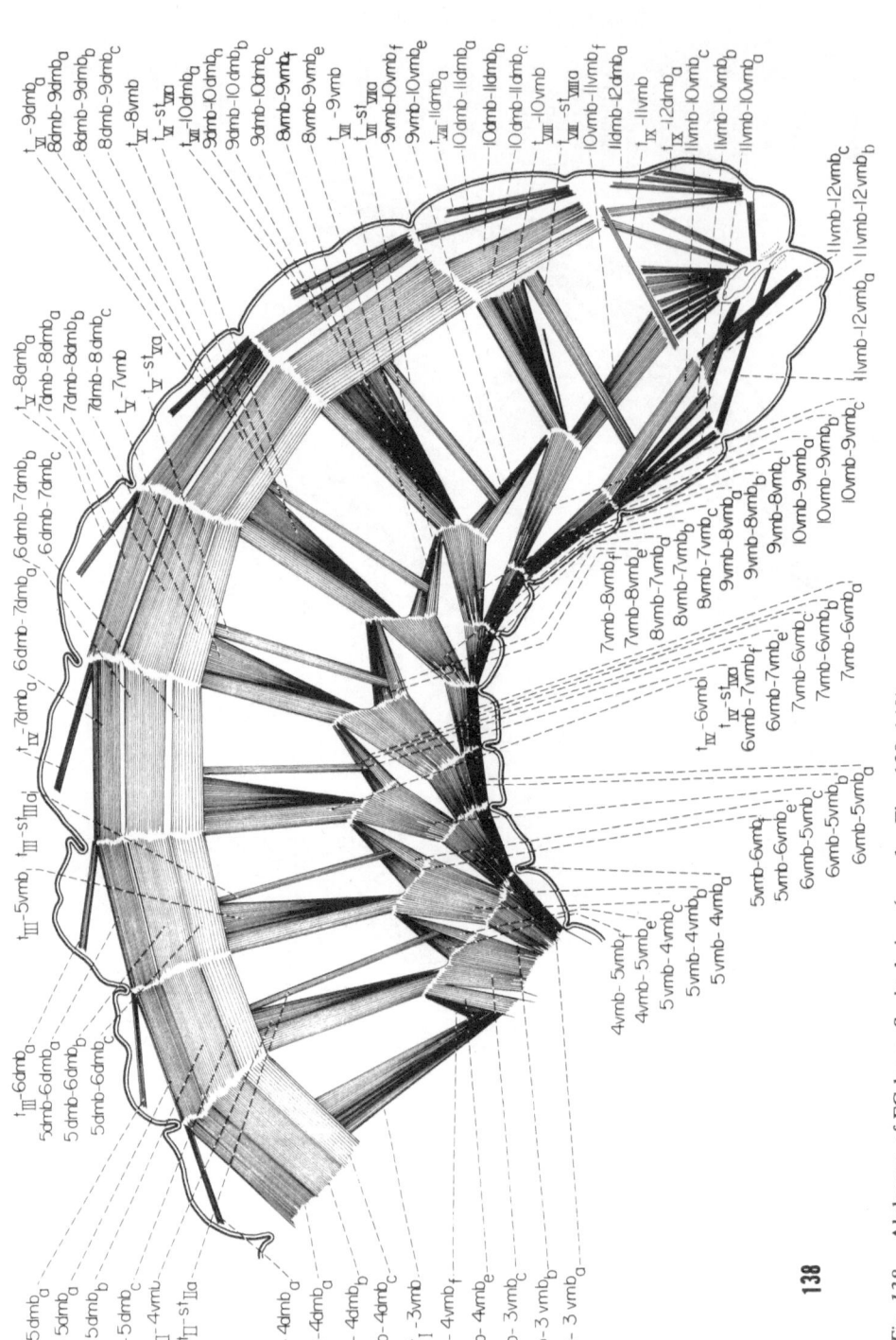

138

Fig. 138. Abdomen of FG larva. Sagittal view (see also Figs. 139–140).

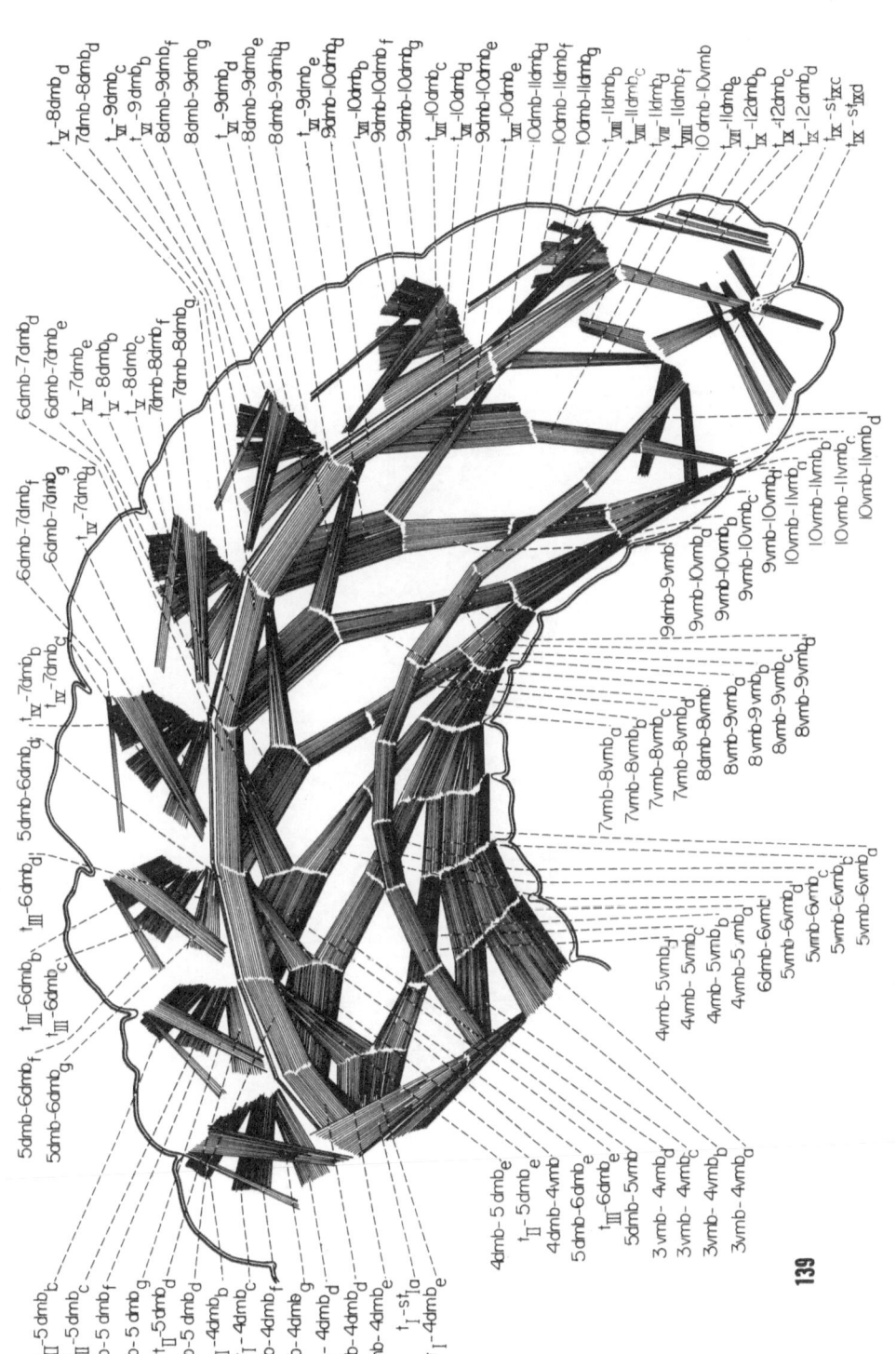

Fig. 139. Abdomen of FG larva. Sagittal view (series continued).

139

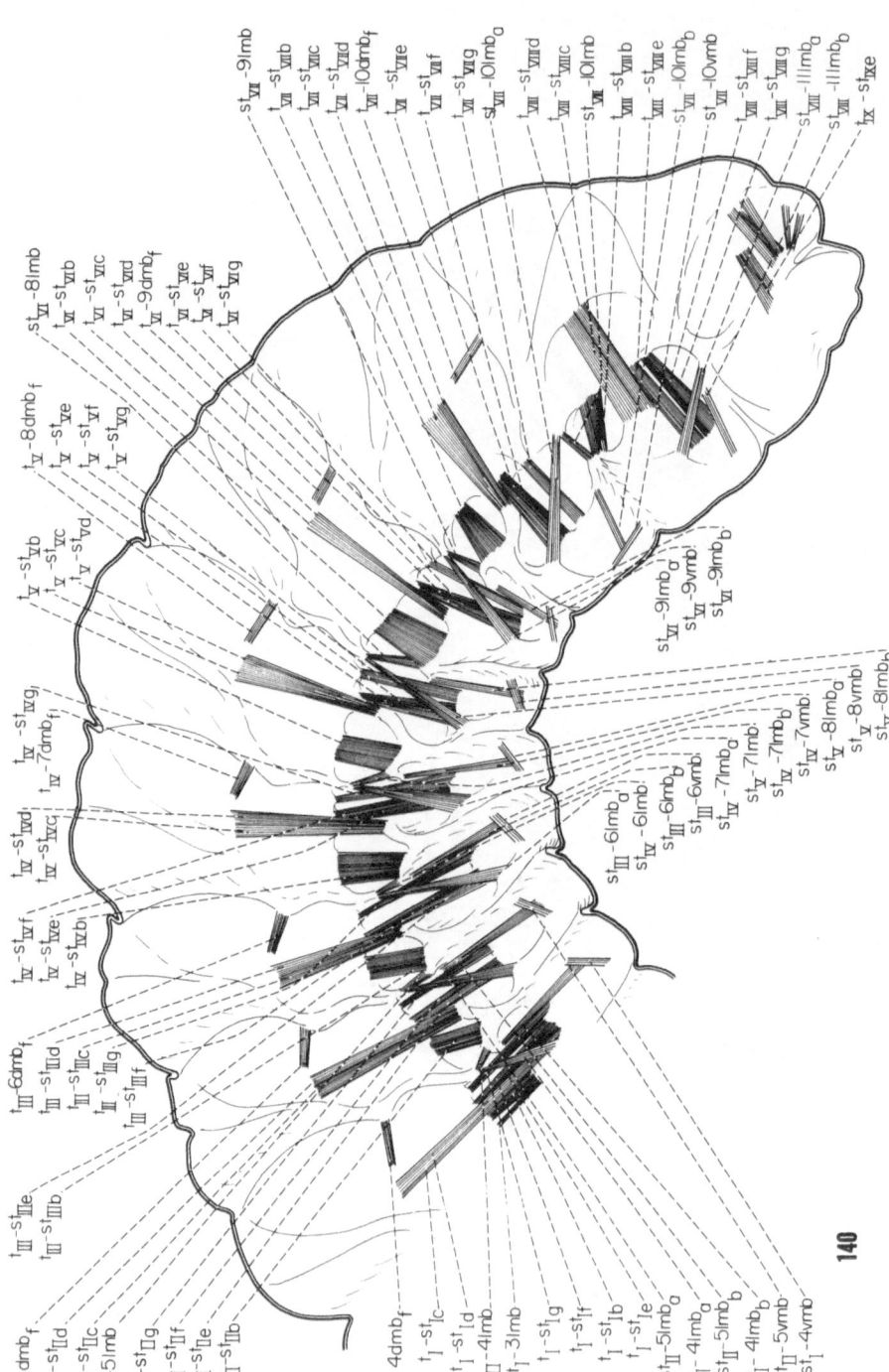

Fig. 140. Abdomen of FG larva. Sagittal view (series concluded).

140

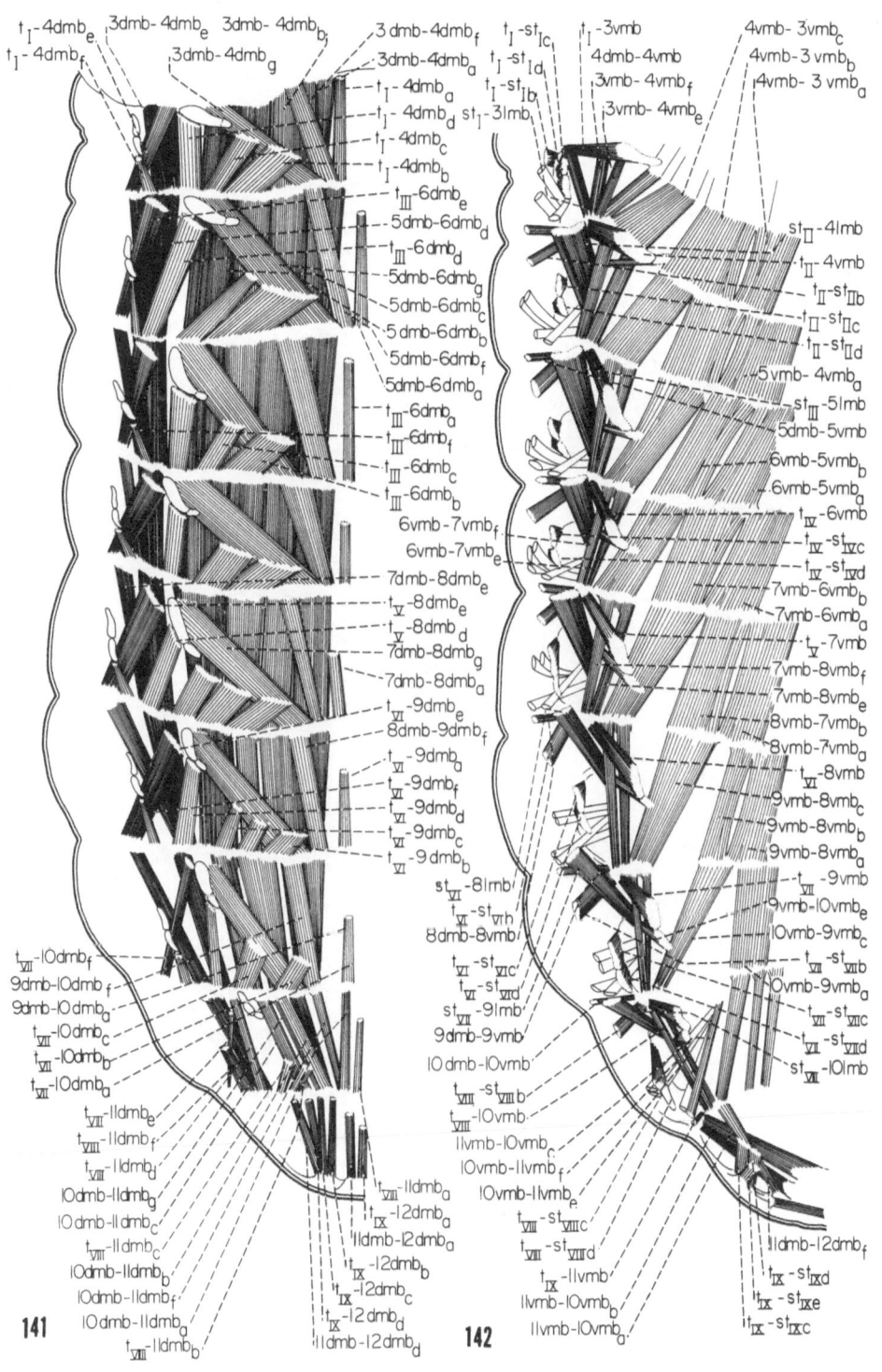

Figs. 141–142. Abdomen of FG larva. Dorsal view (left half) (see also Figs. 143–144).

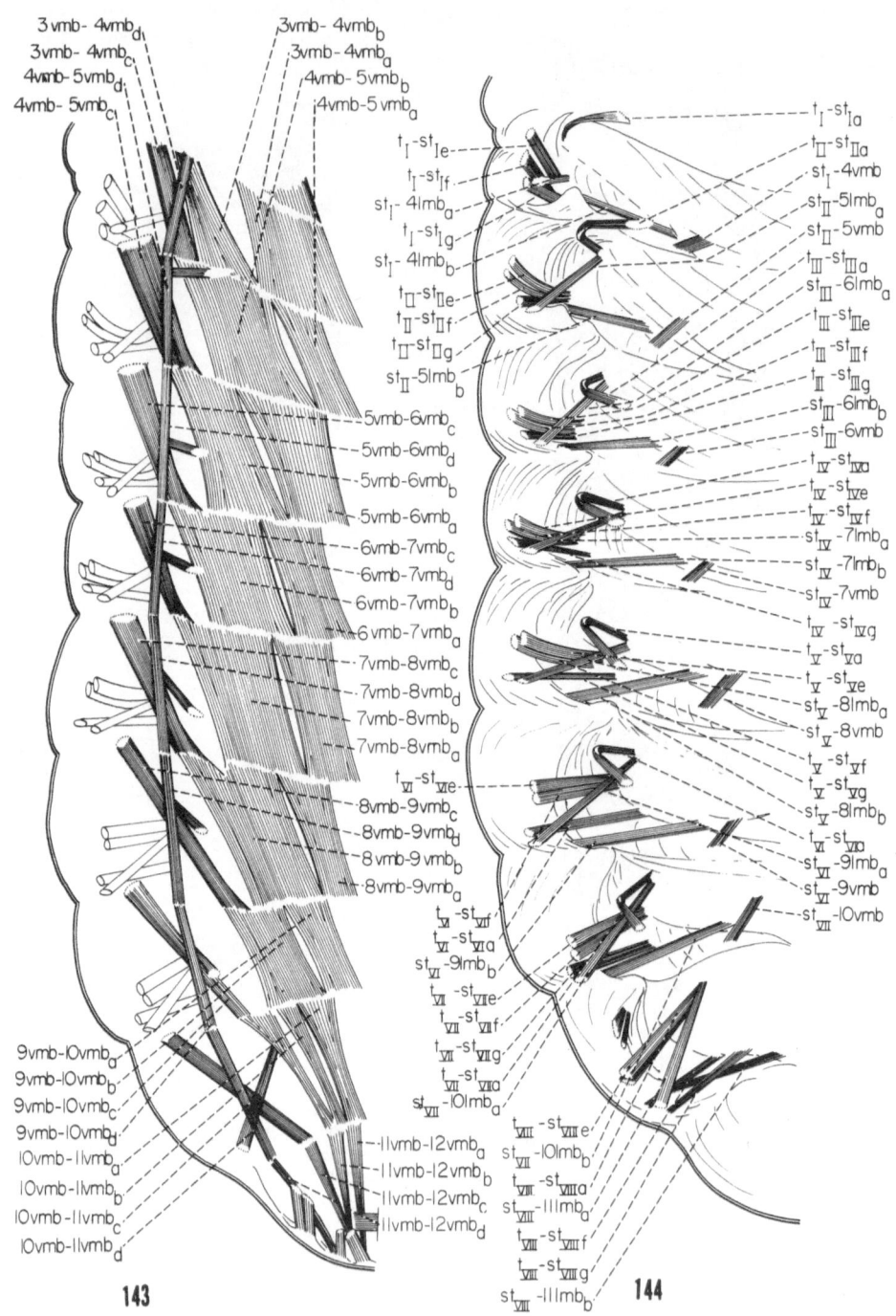

Figs. 143–144. Abdomen of FG larva. Dorsal view (left half) (series concluded).

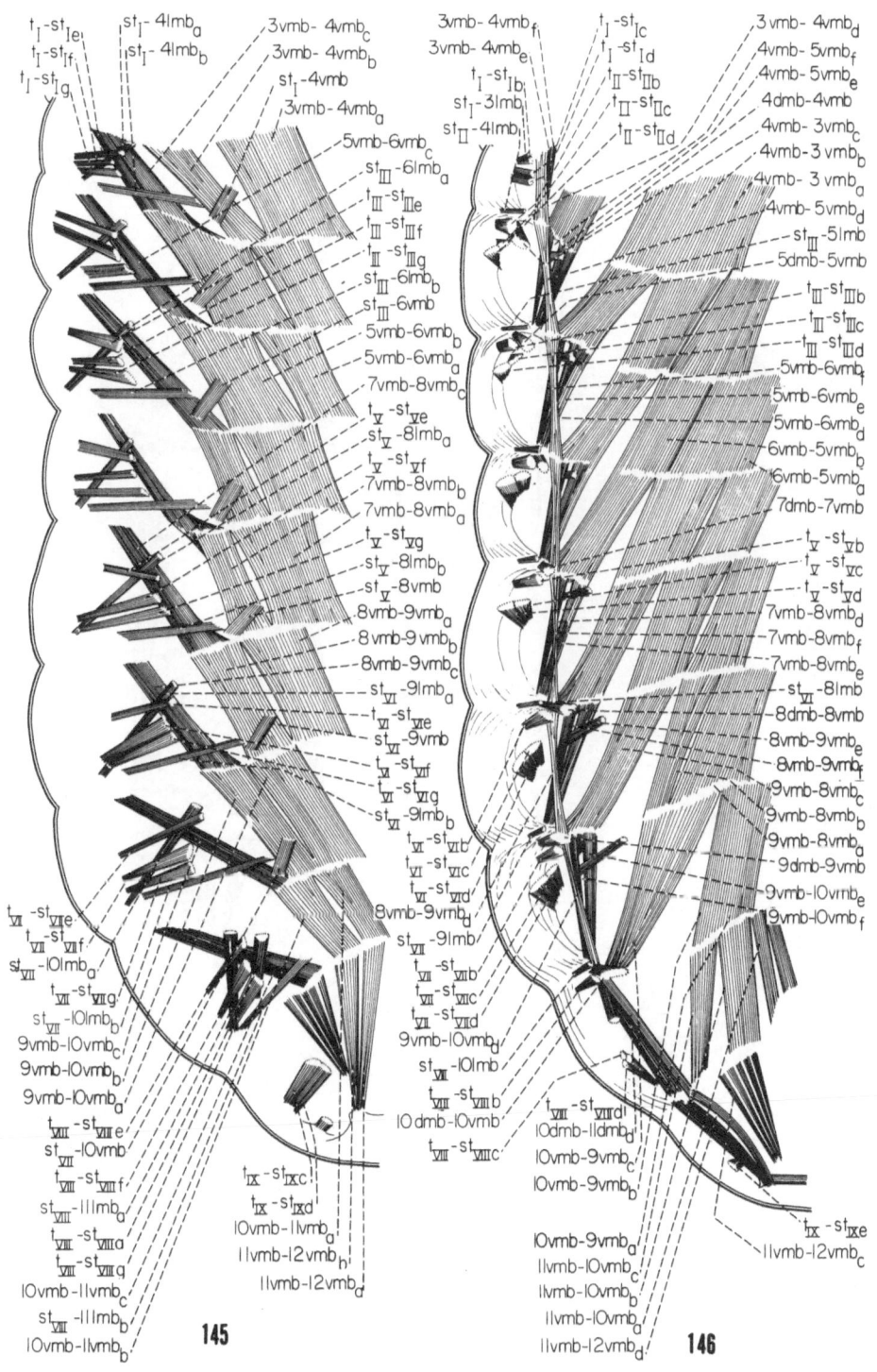

Figs. 145–146. Abdomen of FG larva. Ventral view (right half) (see also Figs. 147–148).

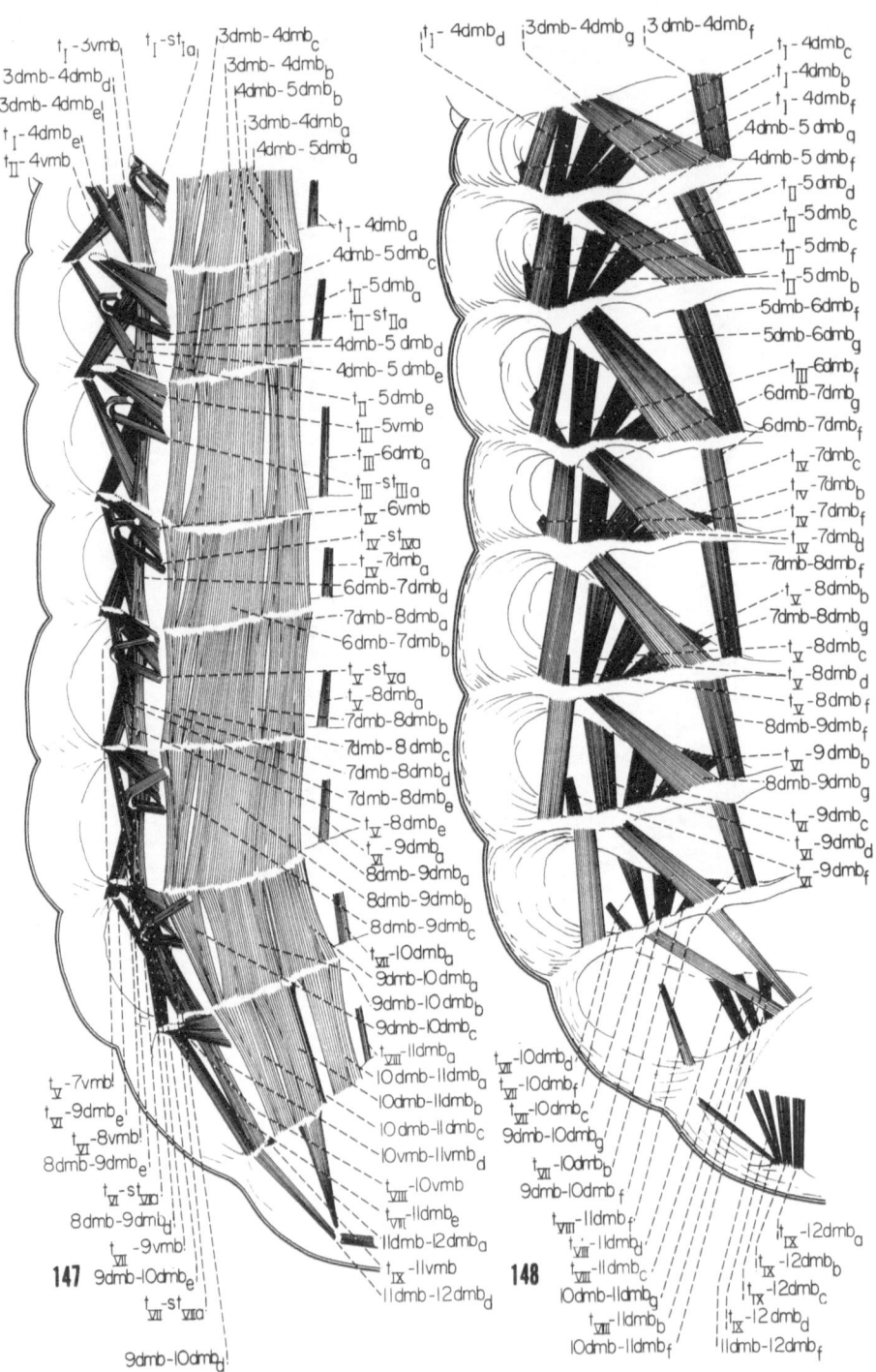

Figs. 147–148. Abdomen of FG larva. Ventral view (right half) (series concluded).

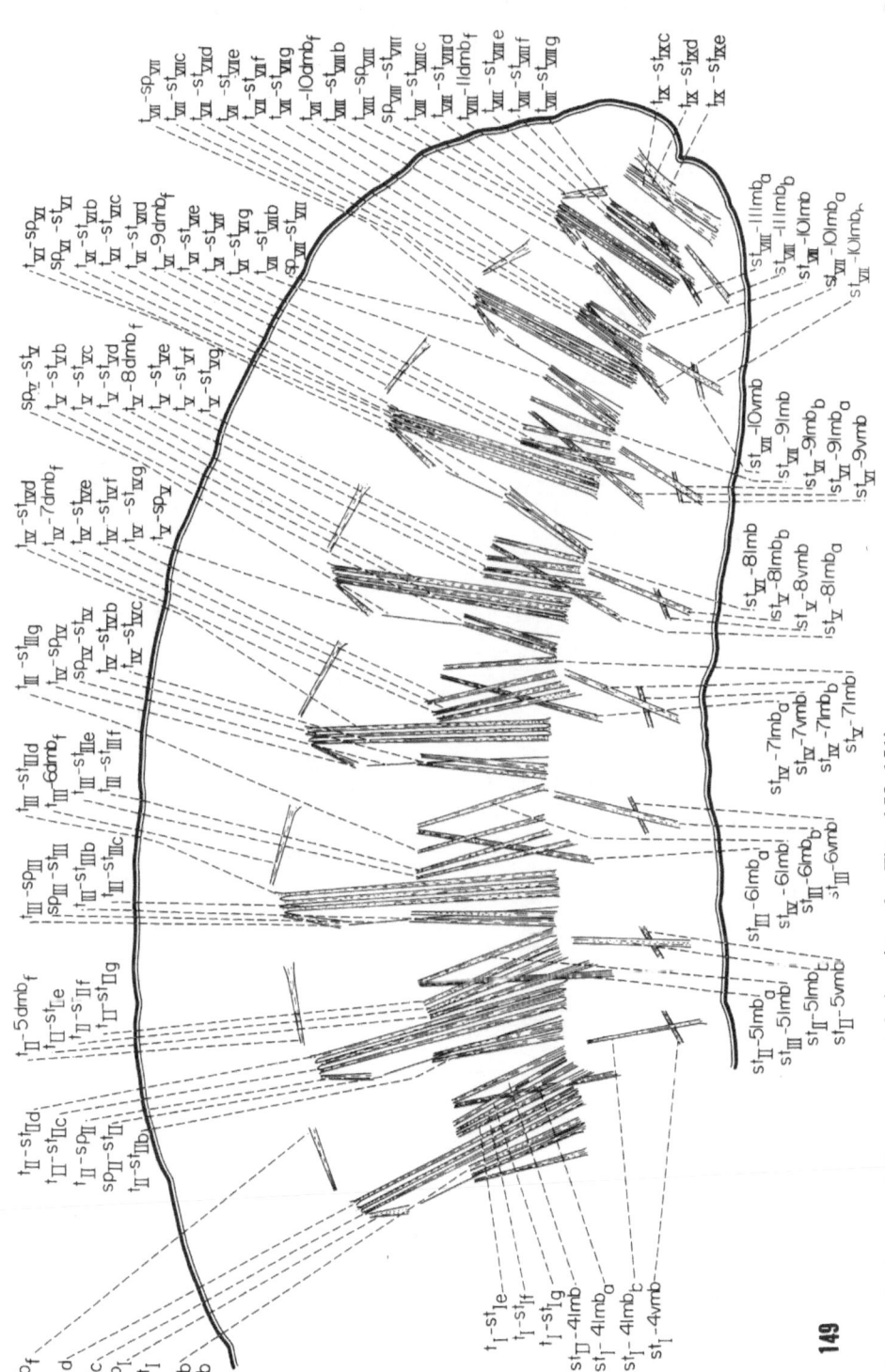

149

Fig. 149. Abdomen of C larva. Lateral view (see also Figs. 150–151).

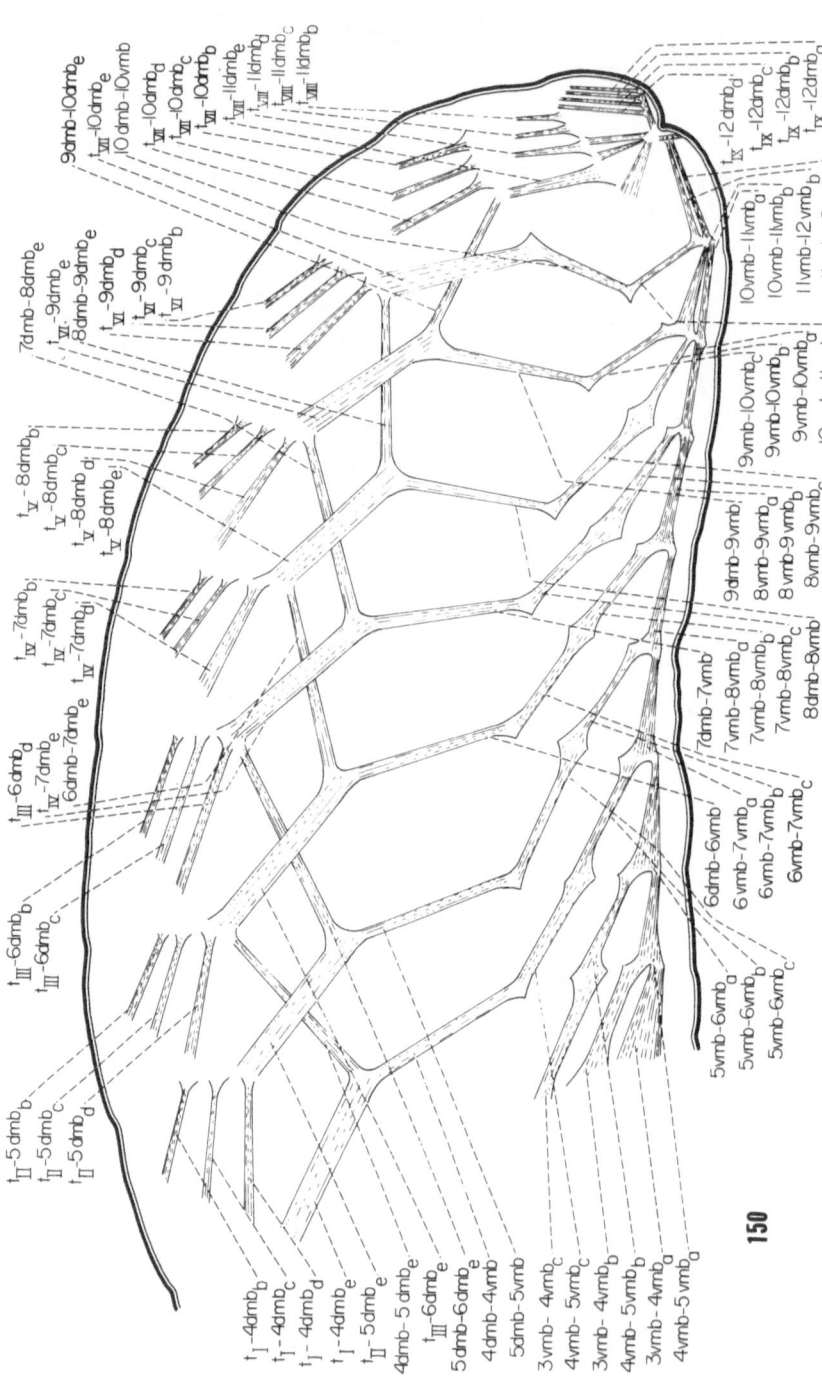

Fig. 150. Abdomen of C larva. Lateral view (series continued).

150

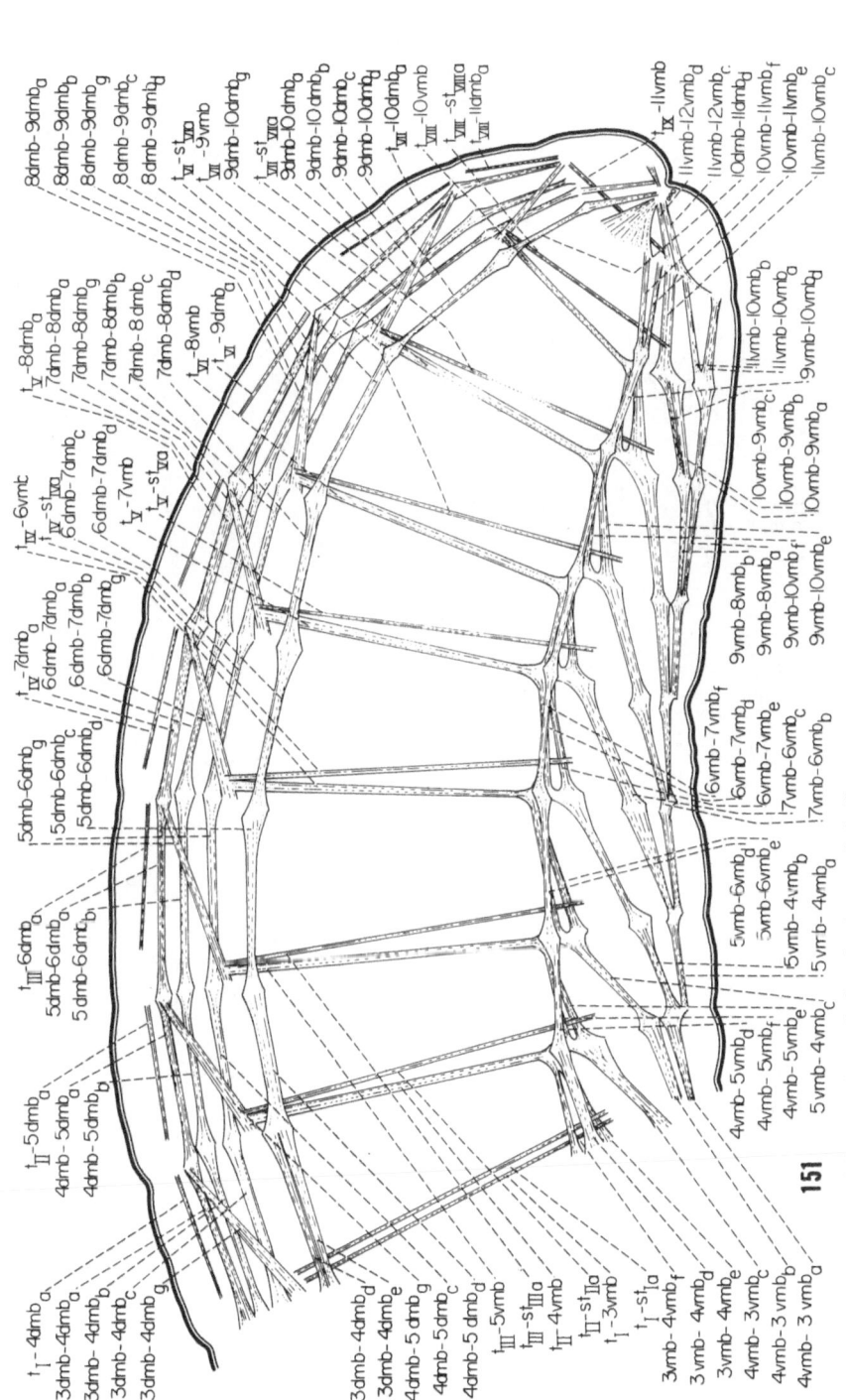

Fig. 151. Abdomen of C larva. Lateral view (series concluded).

151

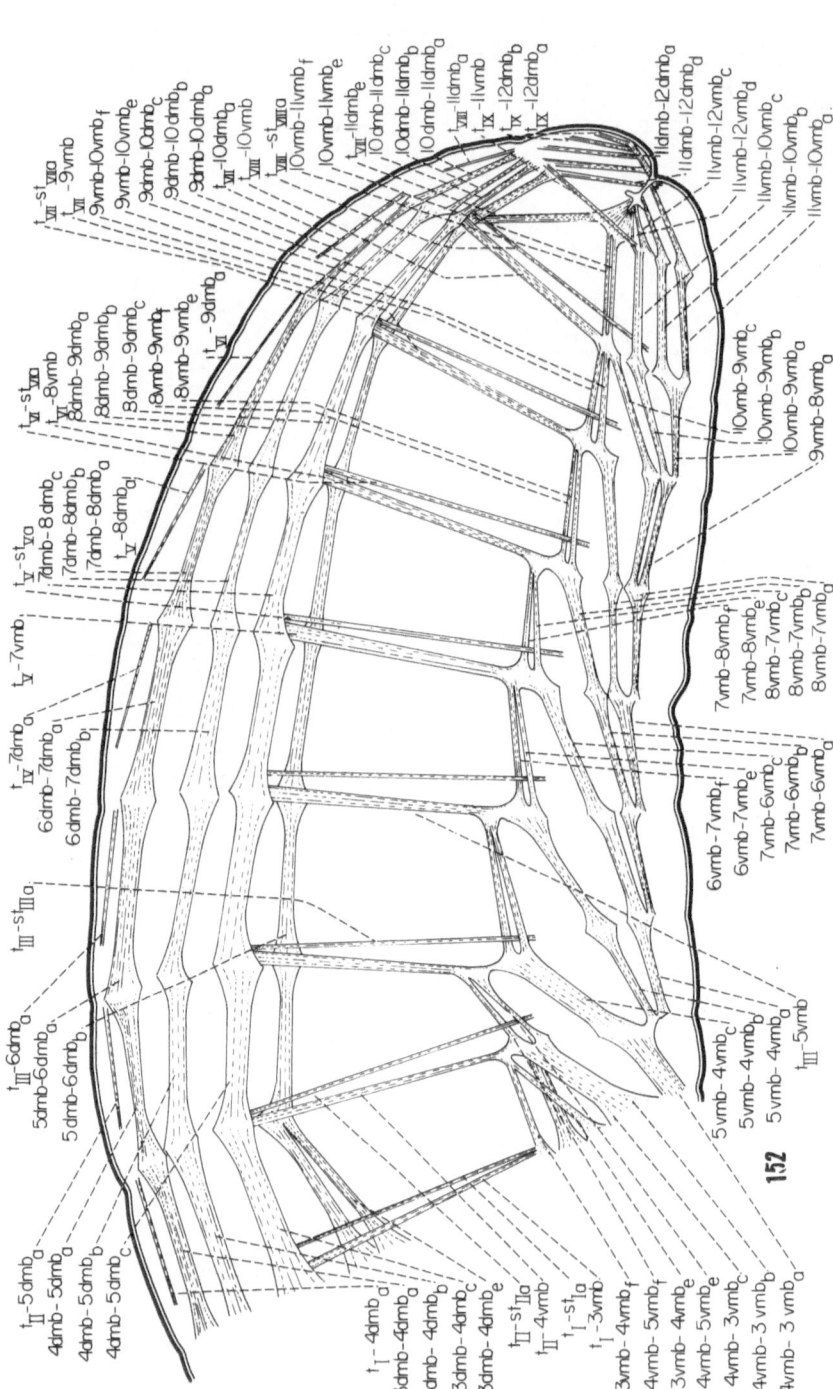

Fig. 152. Abdomen of C larva. Sagittal view (see also Figs. 153–154).

152

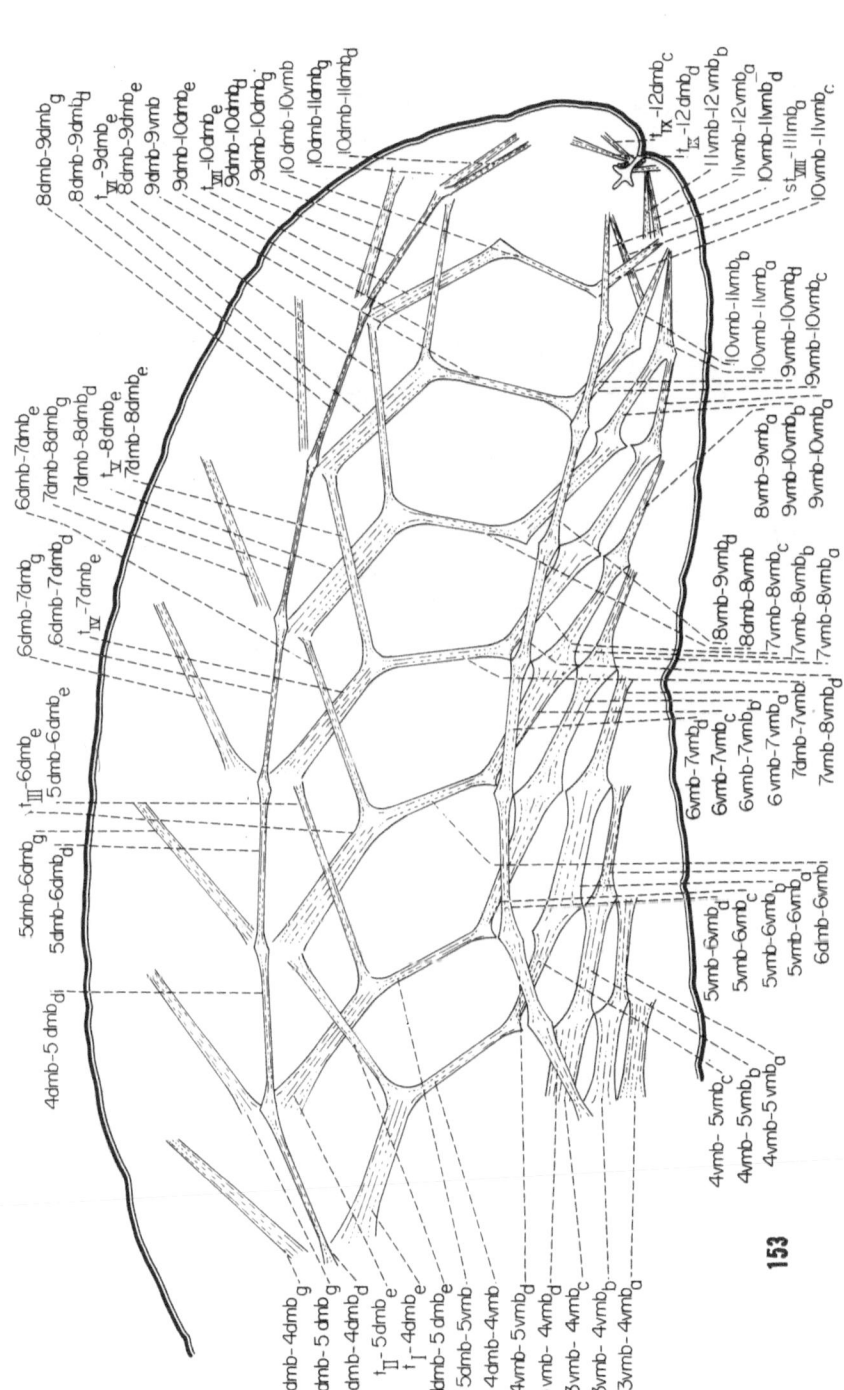

153

Fig. 153. Abdomen of C larva. Sagittal view (series continued).

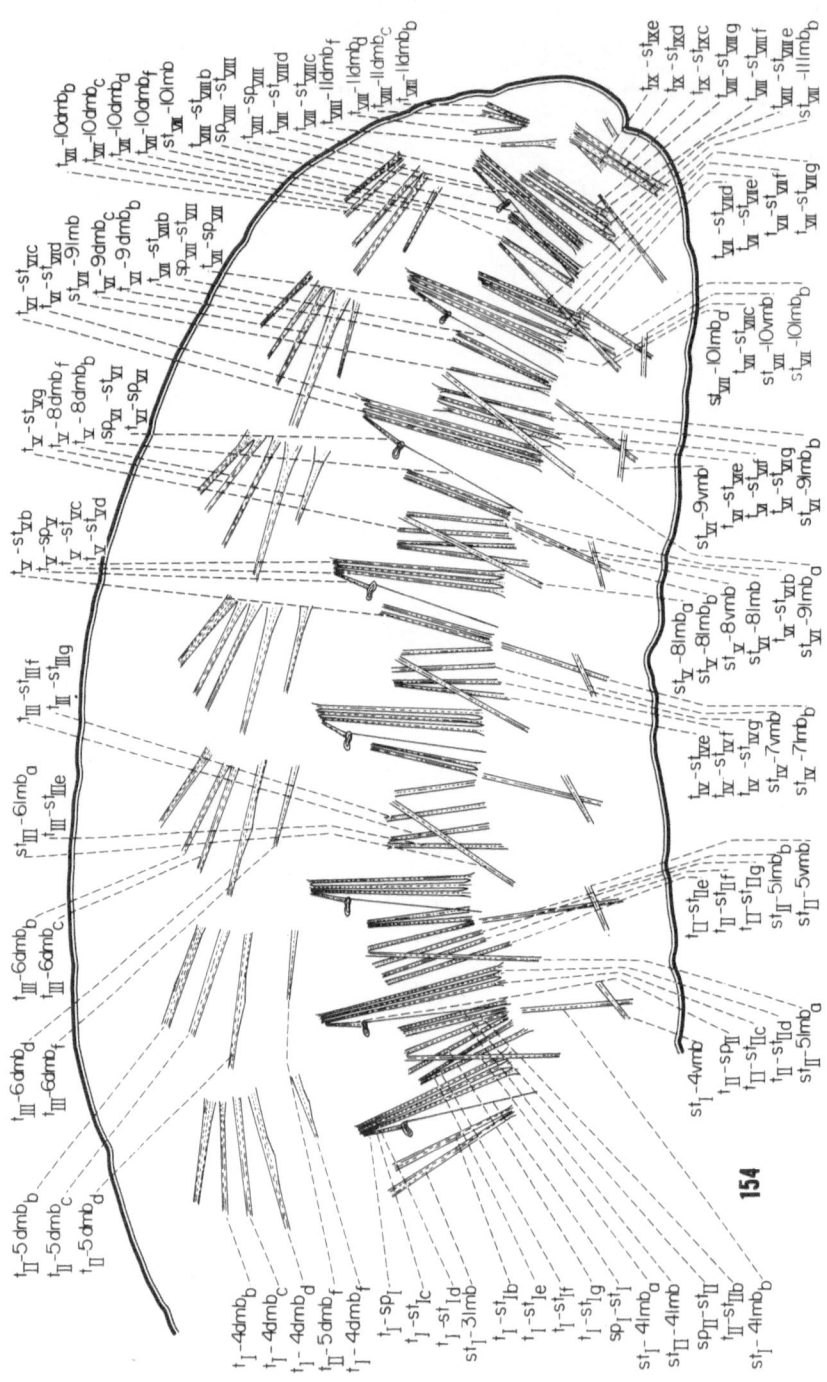

154

Fig. 154. Abdomen of C larva. Sagittal view (series concluded).

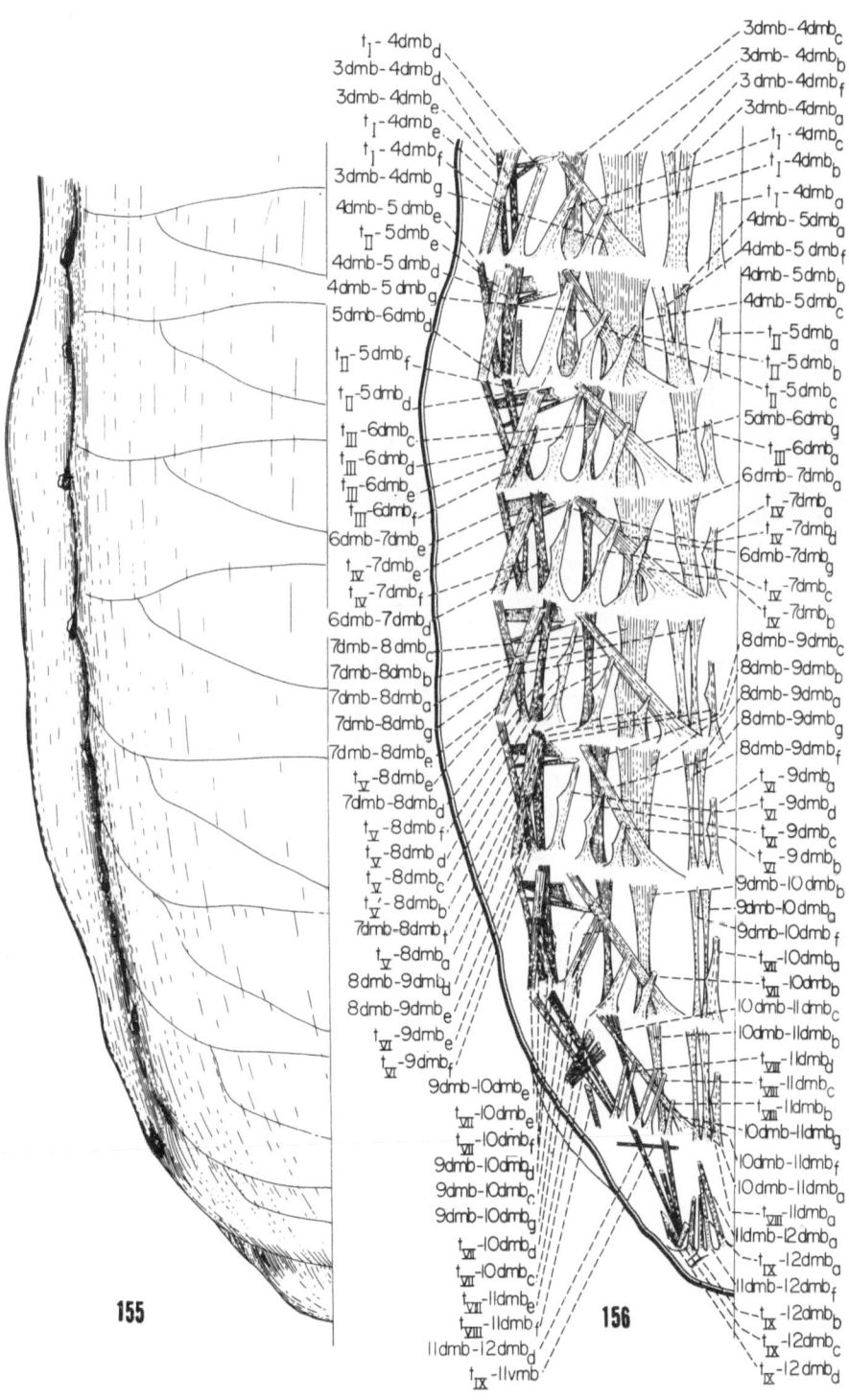

155

156

Figs. 155–156. Abdomen of C larva. Dorsal view (left half) (see also Figs. 157–160).

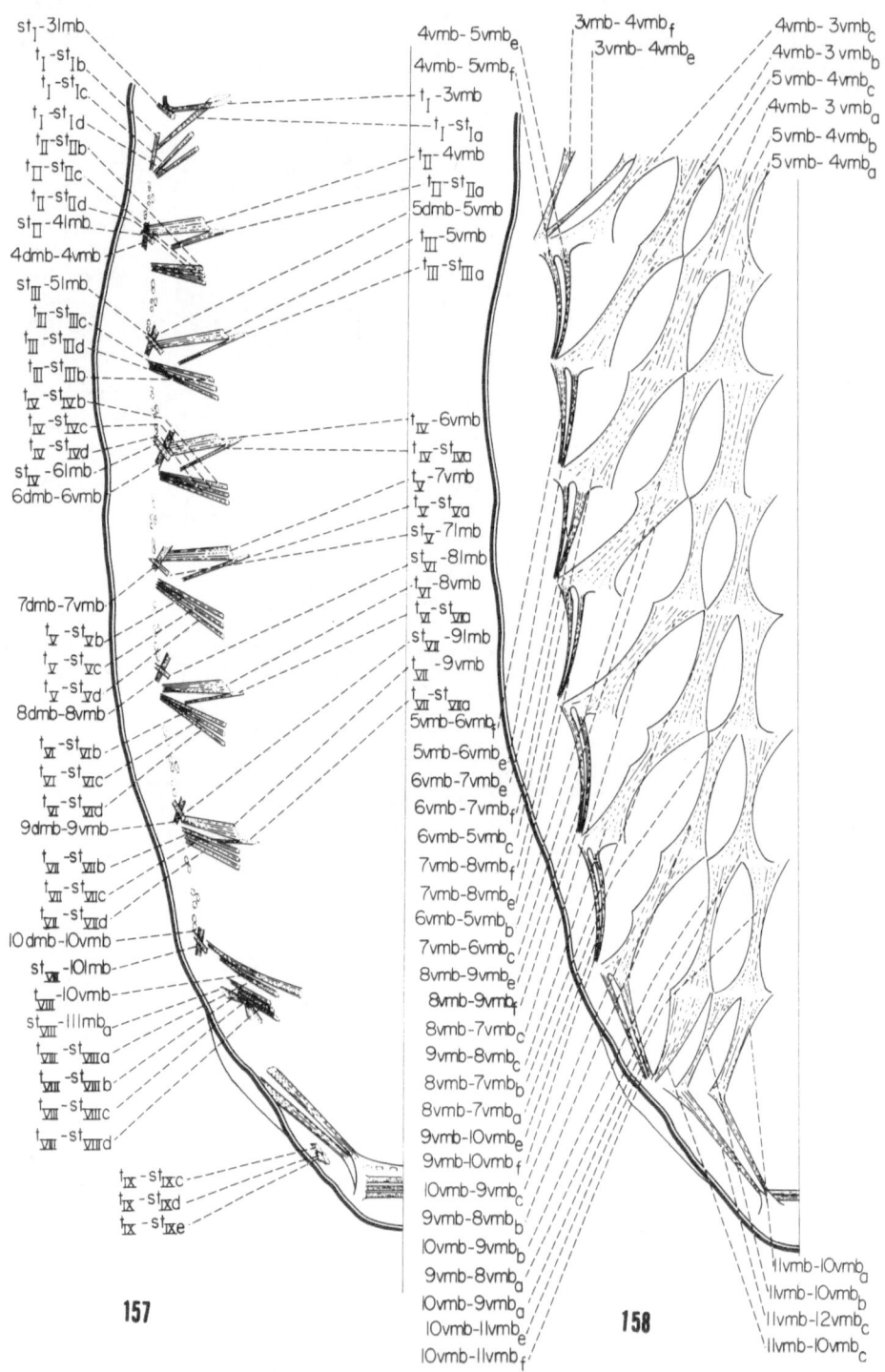

Figs. 157–158. Abdomen of C larva. Dorsal view (left half) (series continued).

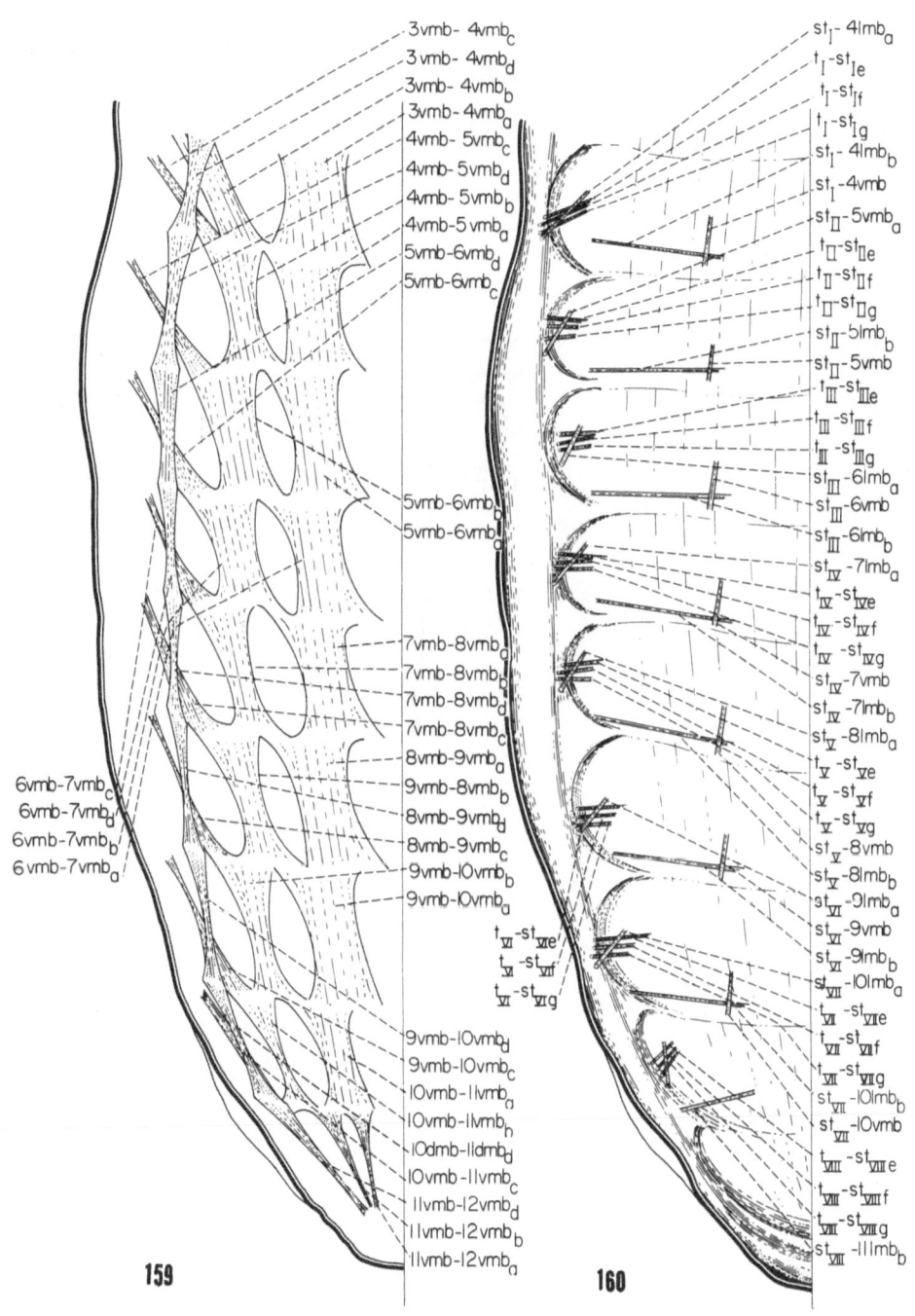

Figs. 159–160. Abdomen of C larva. Dorsal view (left half) (series concluded).

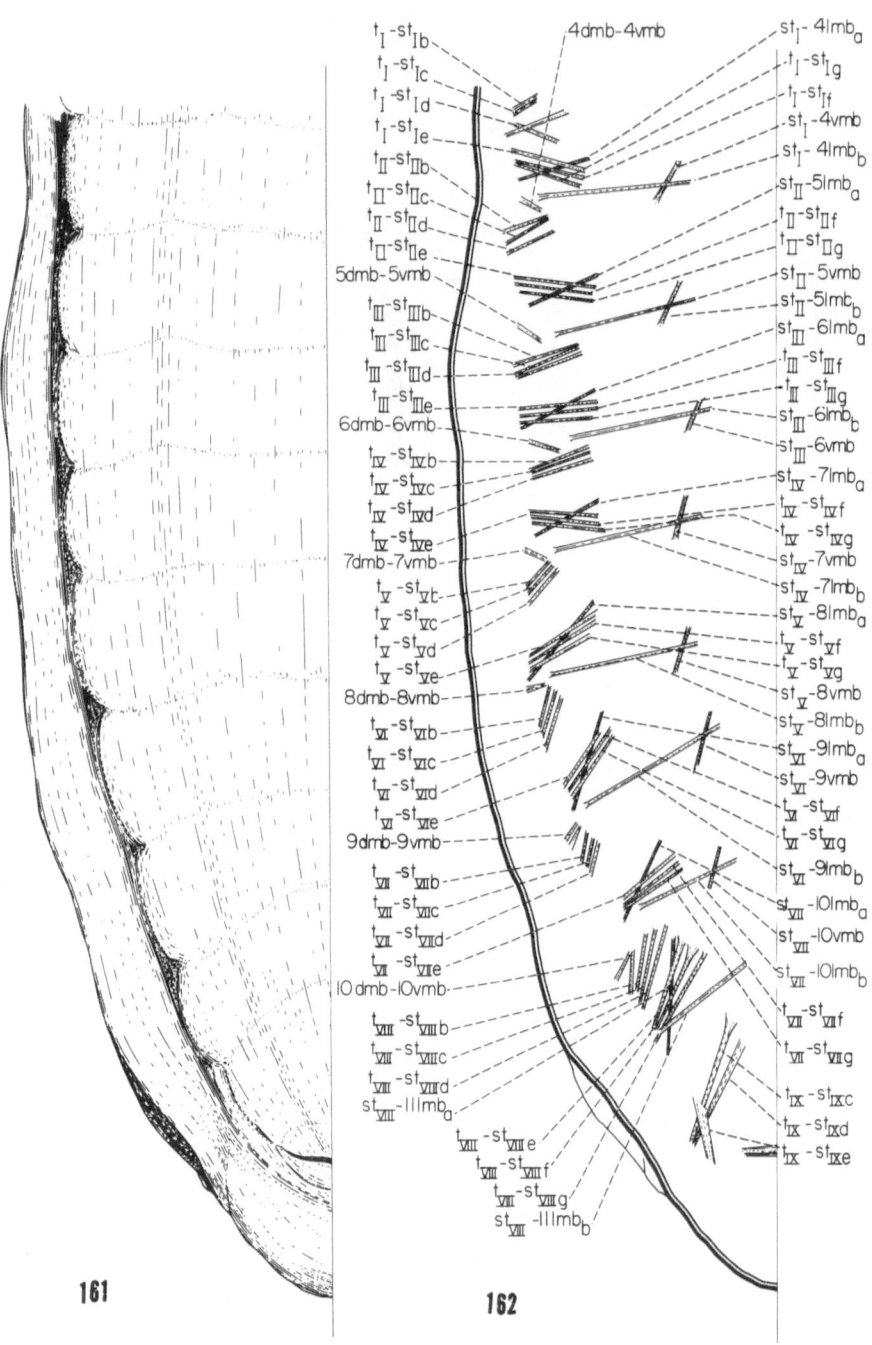

Figs. 161–162. Abdomen of C larva. Ventral view (right half) (see also Figs. 163–166).

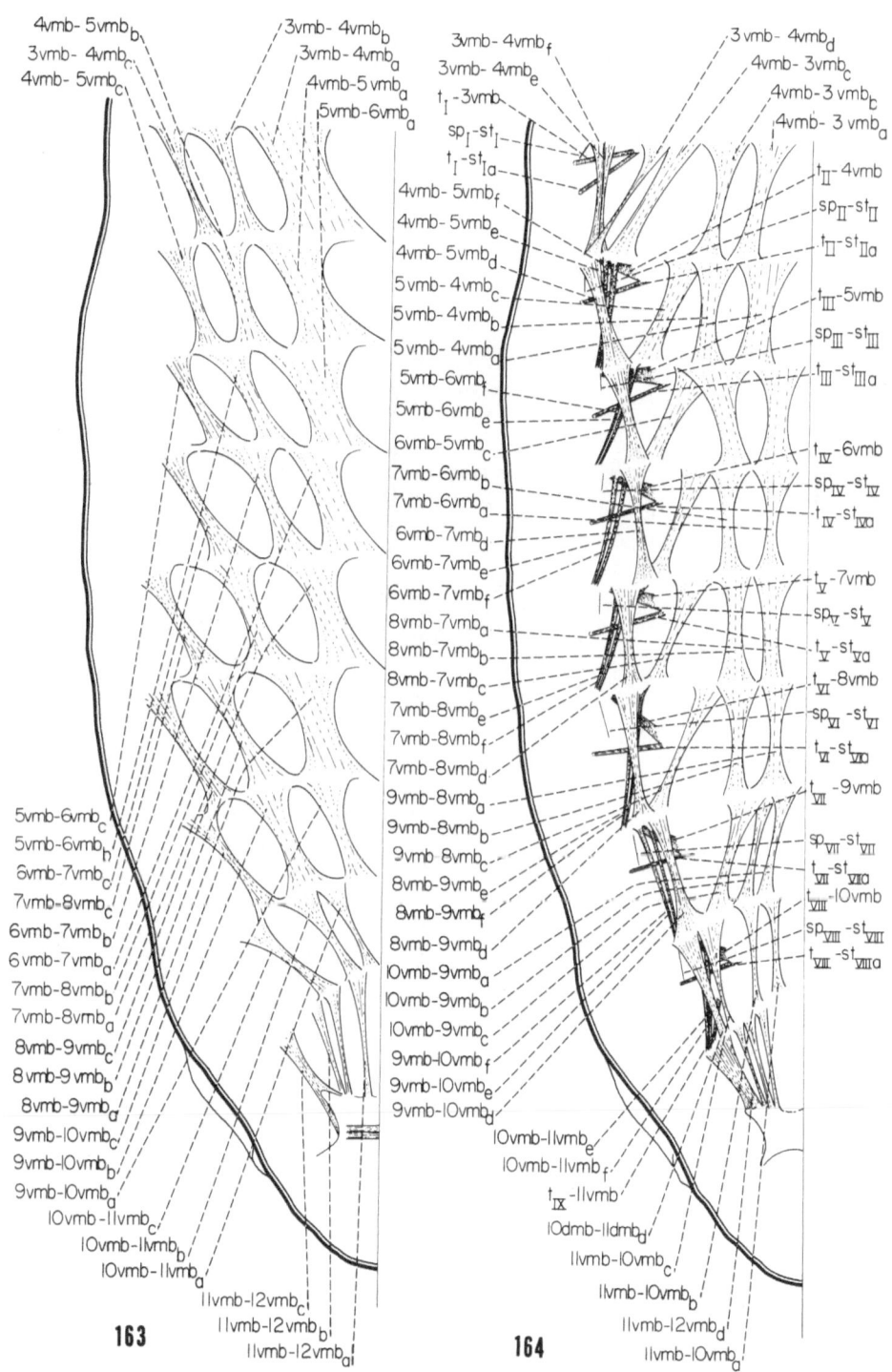

Figs. 163–164. Abdomen of C larva. Ventral view (right half) (series continued).

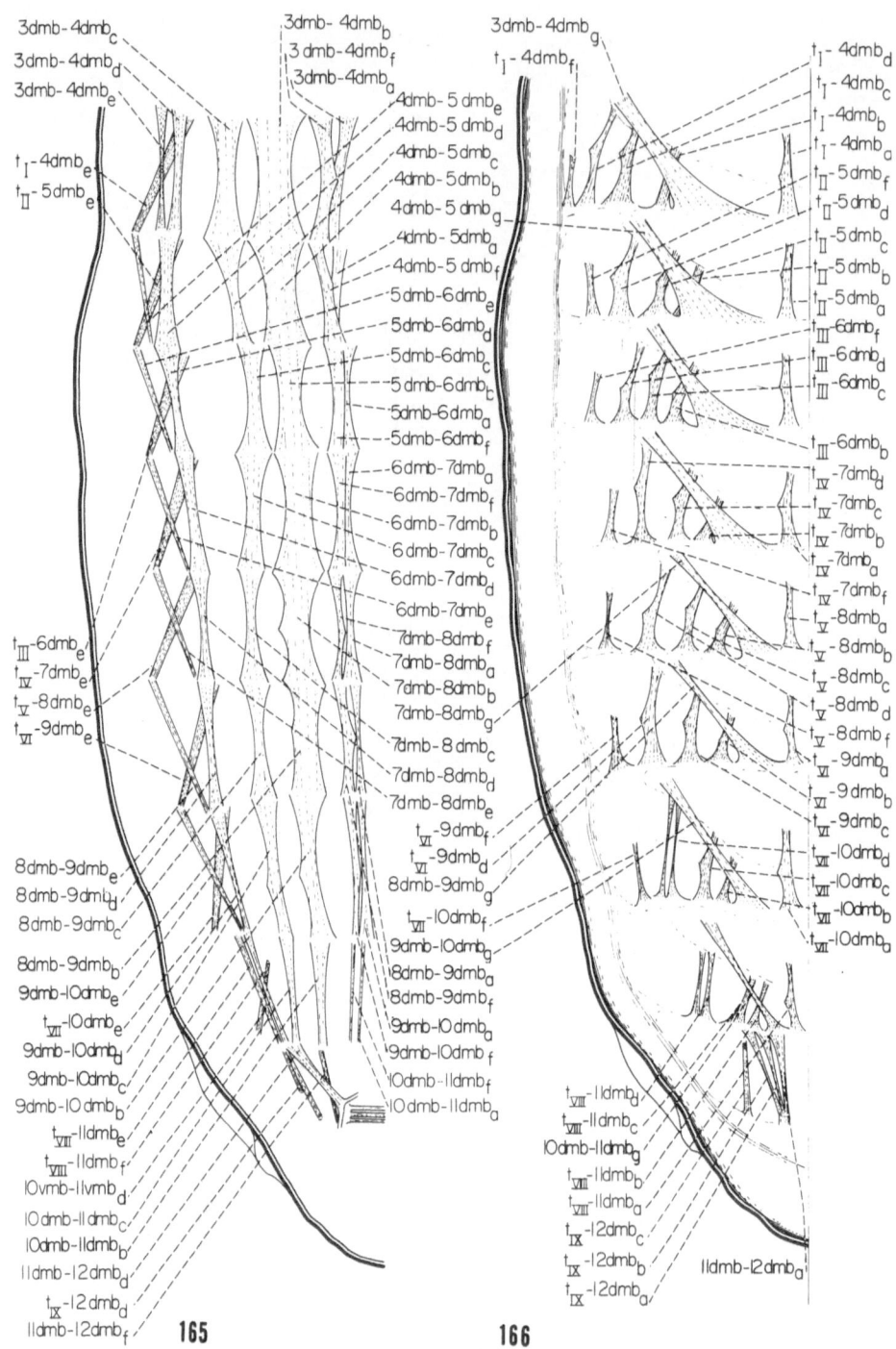

165

166

Figs. 165–166. Abdomen of C larva. Ventral view (right half) (series concluded).

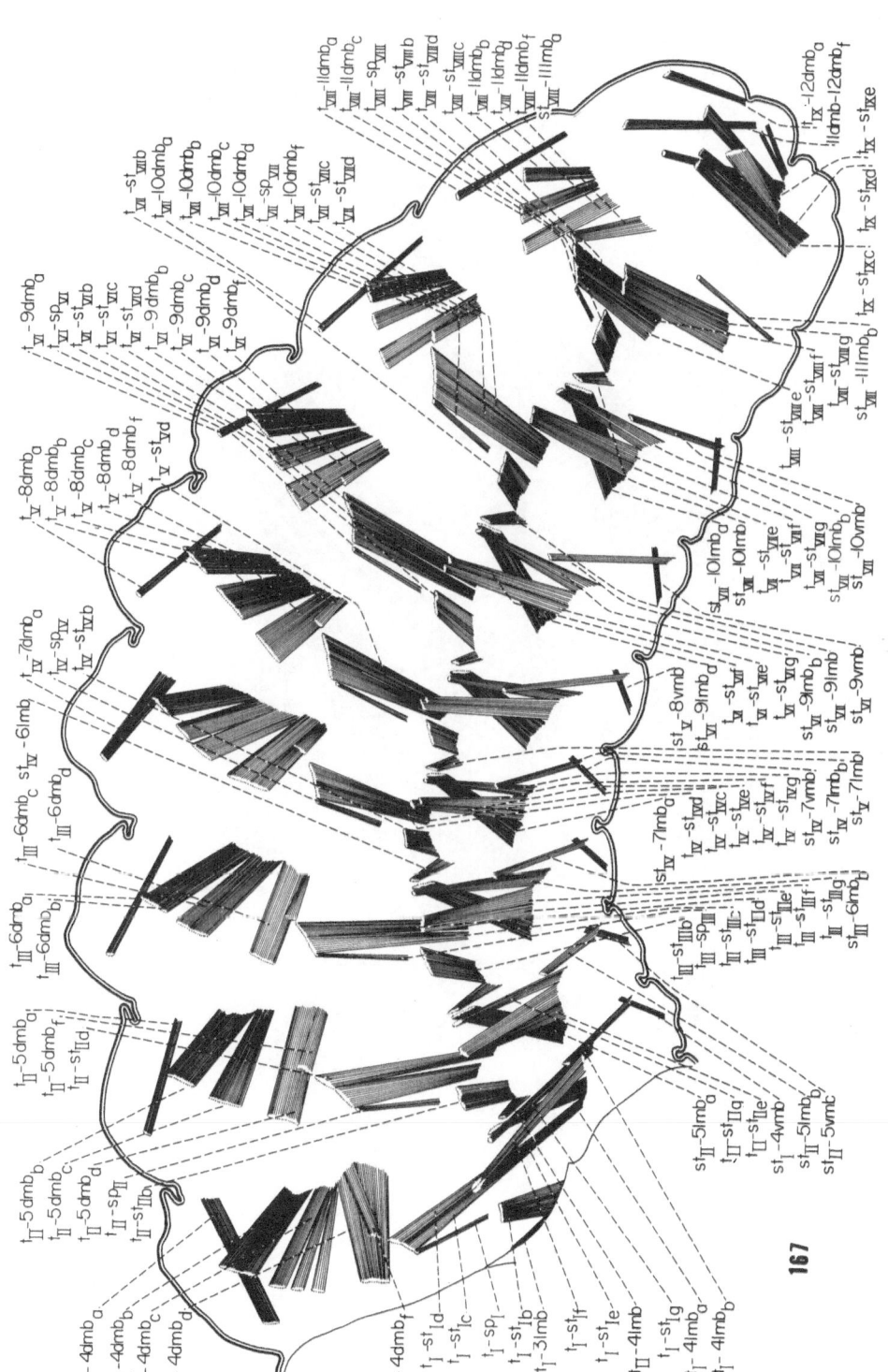

167

Fig. 167. Abdomen of SG larva. Lateral view (see also Figs. 168–169).

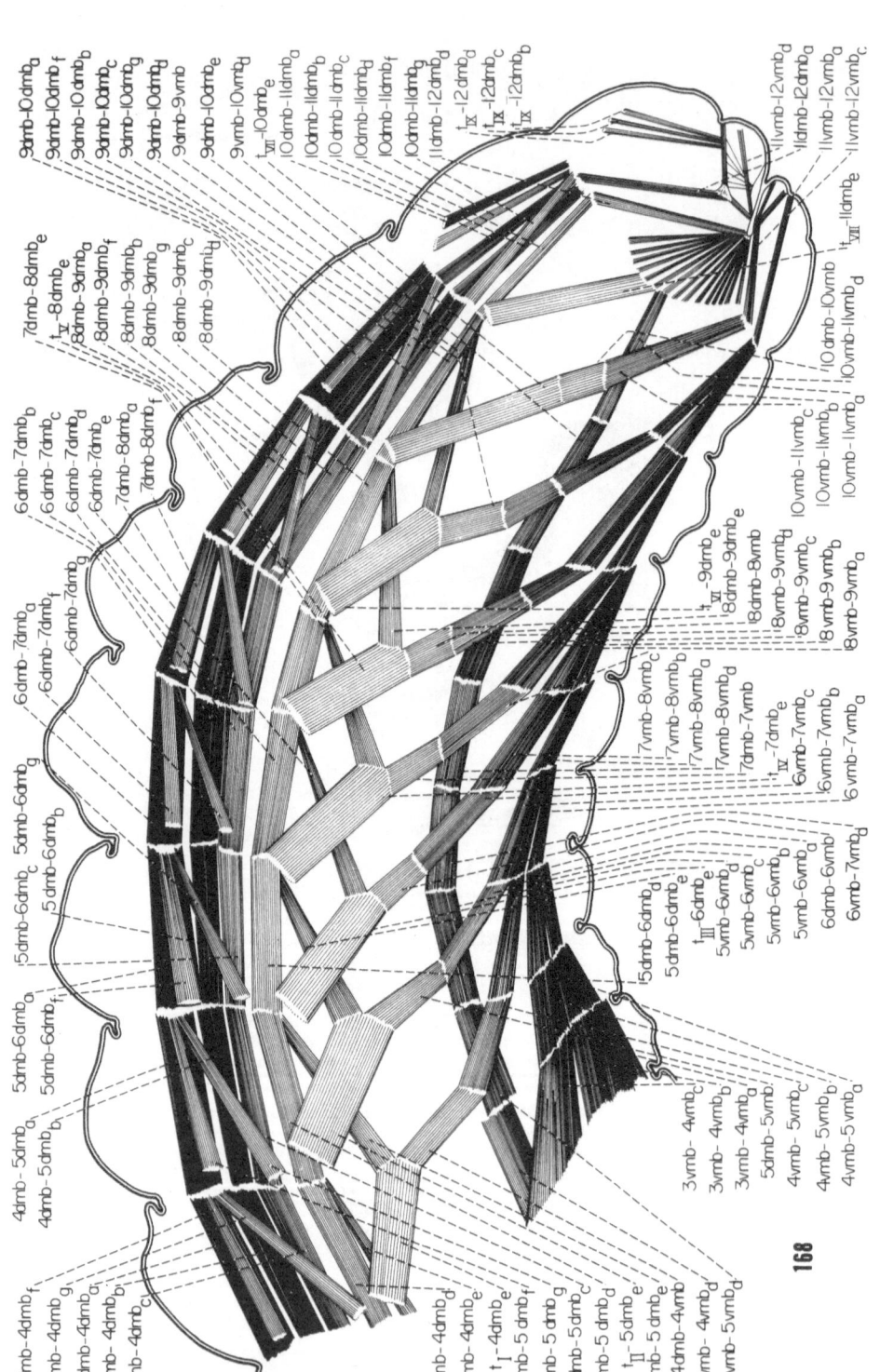

168

Fig. 168. Abdomen of SG larva. Lateral view (series continued)

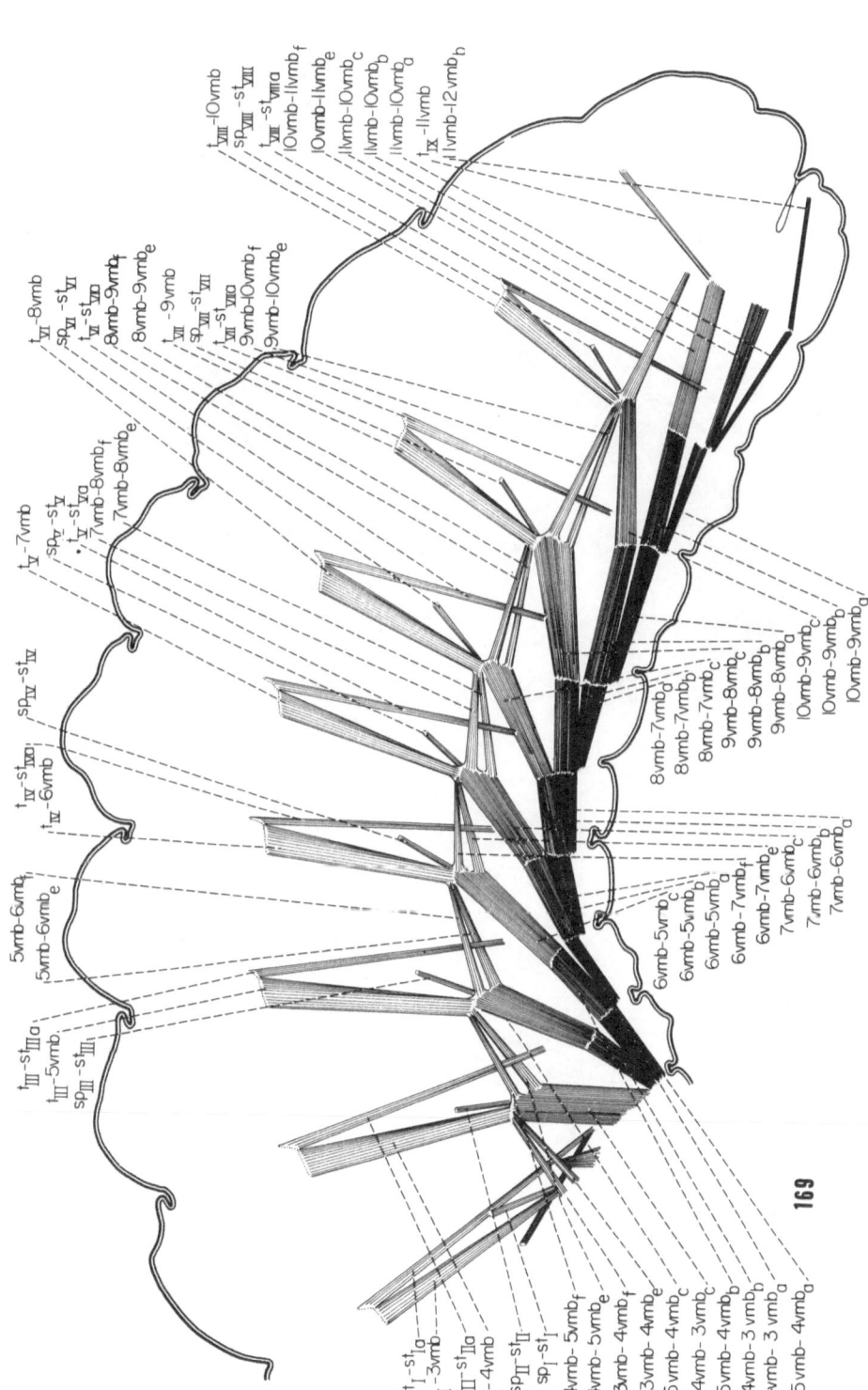

Fig. 169. Abdomen of SG larva. Lateral view (series concluded).

169

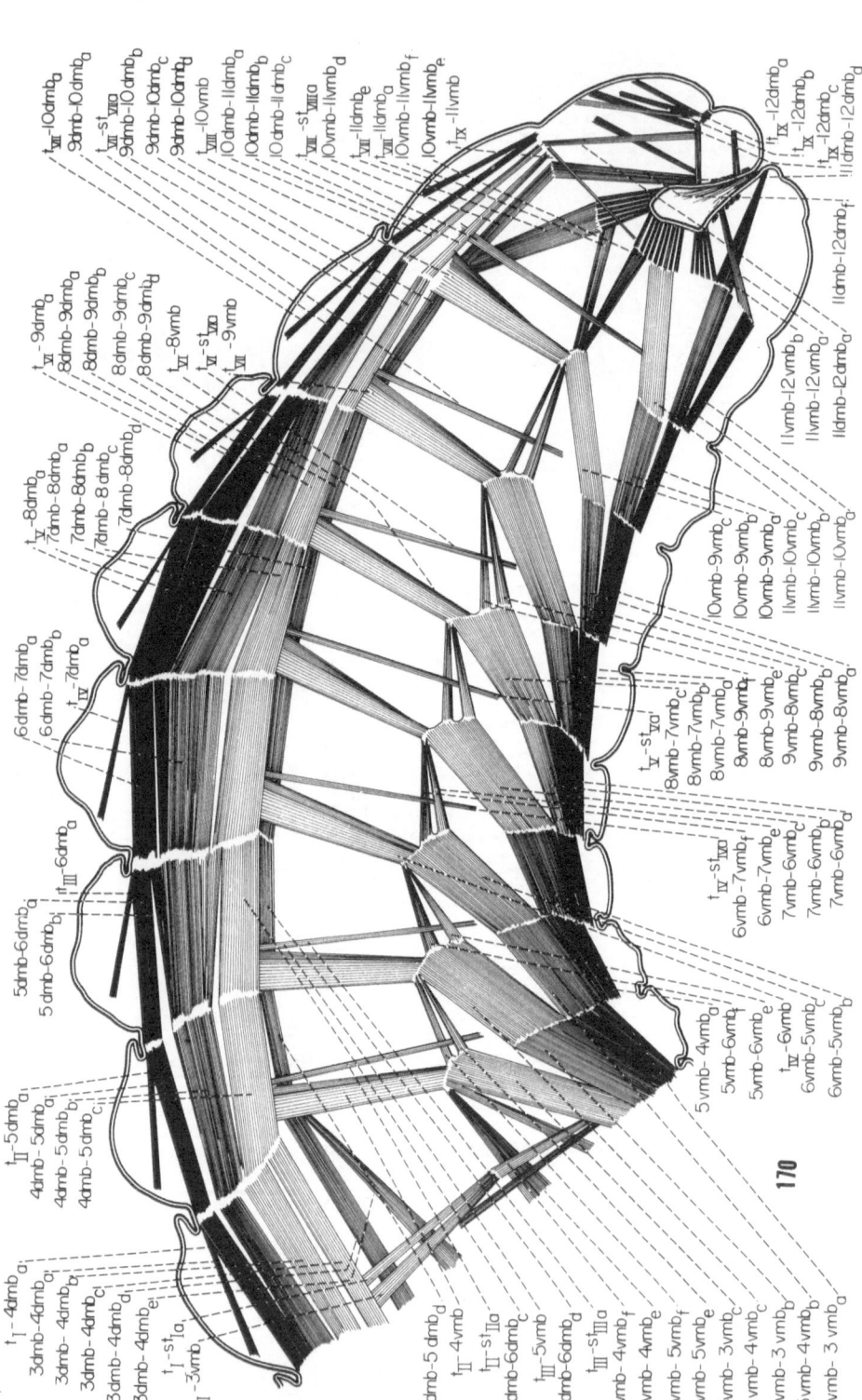

Fig. 170. Abdomen of SG larva. Sagittal view (see also Figs. 171–172).

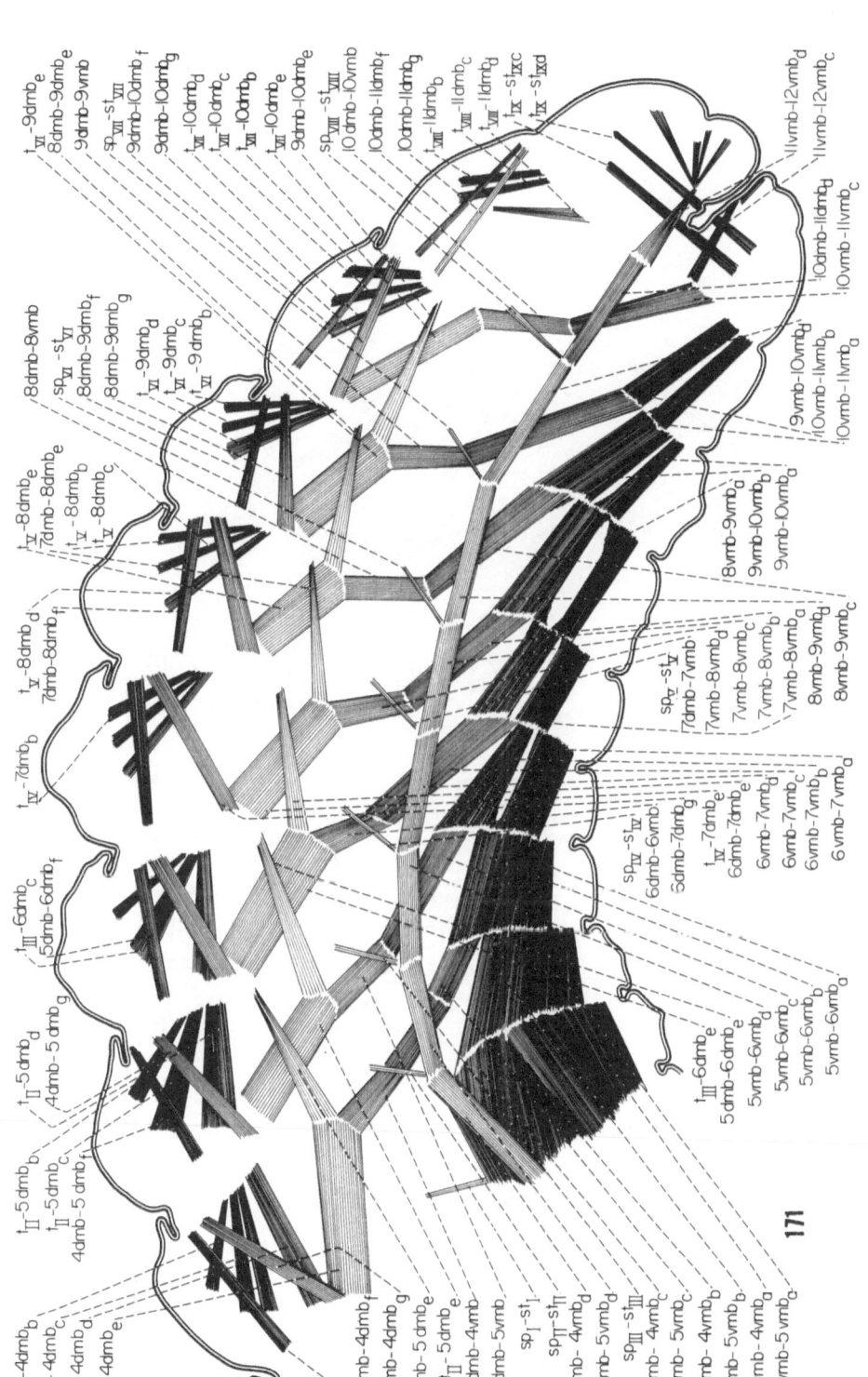

Fig. 171. Abdomen of SG larva. Sagittal view (series continued).

171

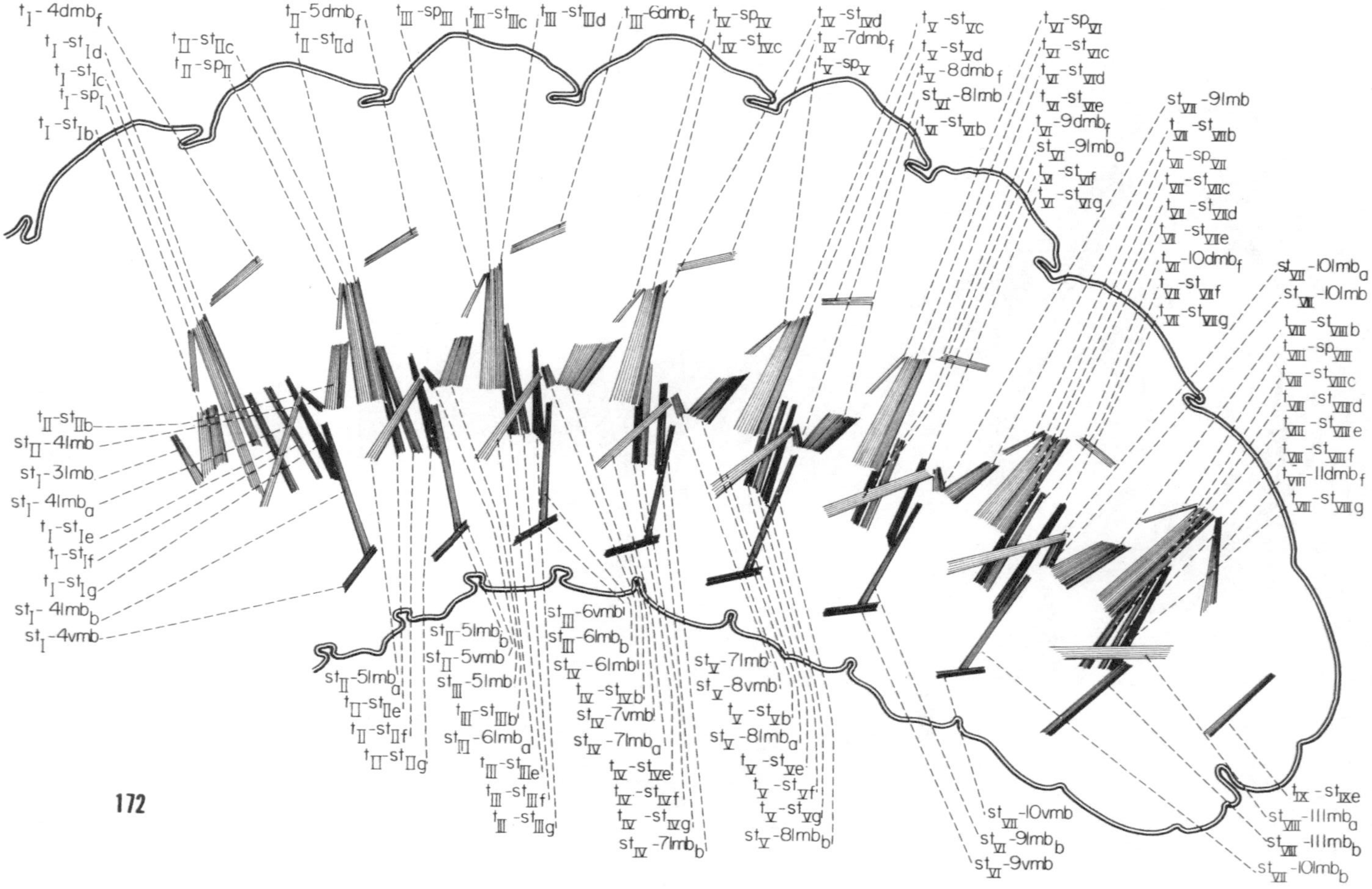

Fig. 172. Abdomen of SG larva. Sagittal view (series concluded).

172

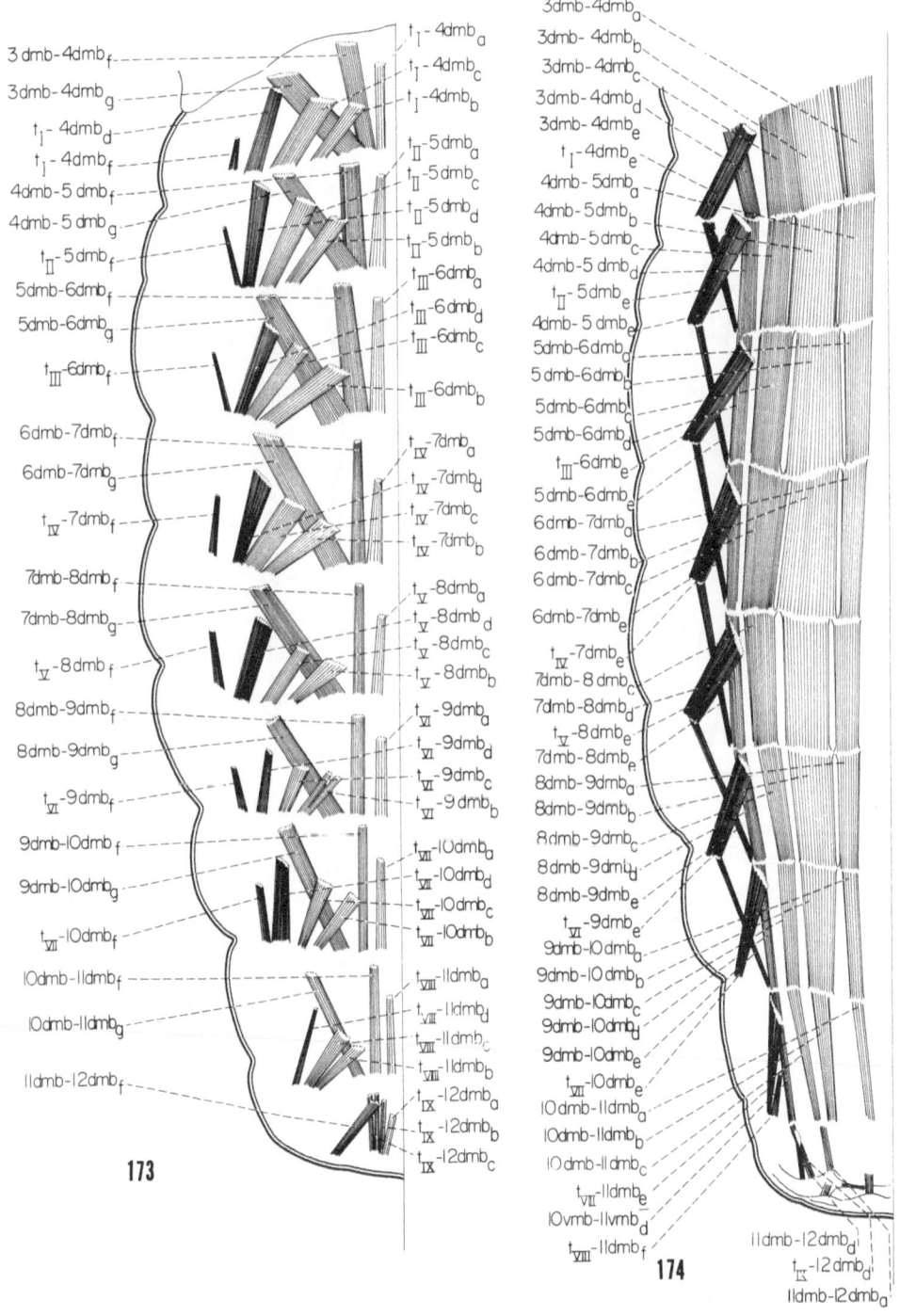

Figs. 173–174.   Abdomen of SG larva. Dorsal view (left half) (see also Figs. 175–178).

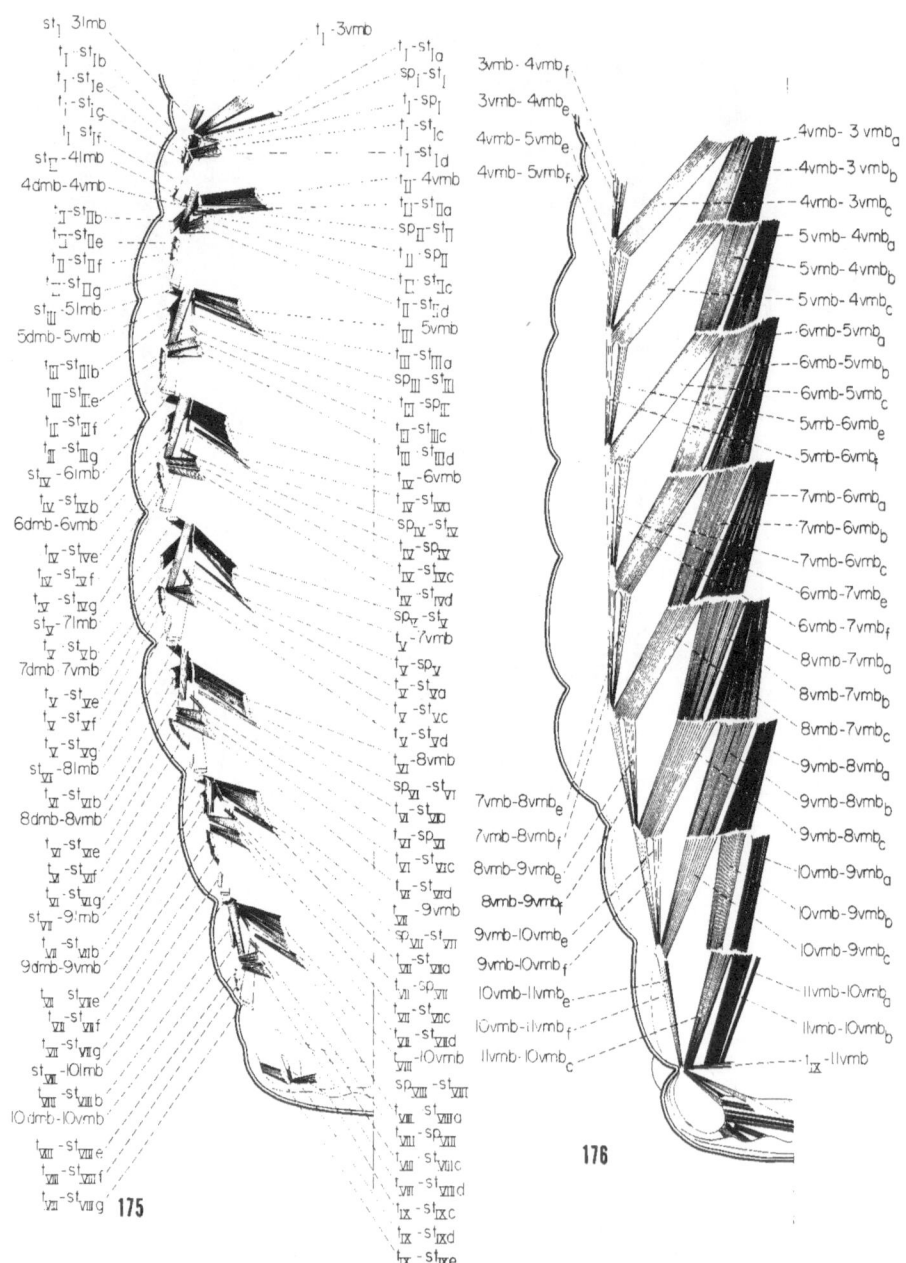

Figs. 175–176. Abdomen of SG larva. Dorsal view (left half) (series continued).

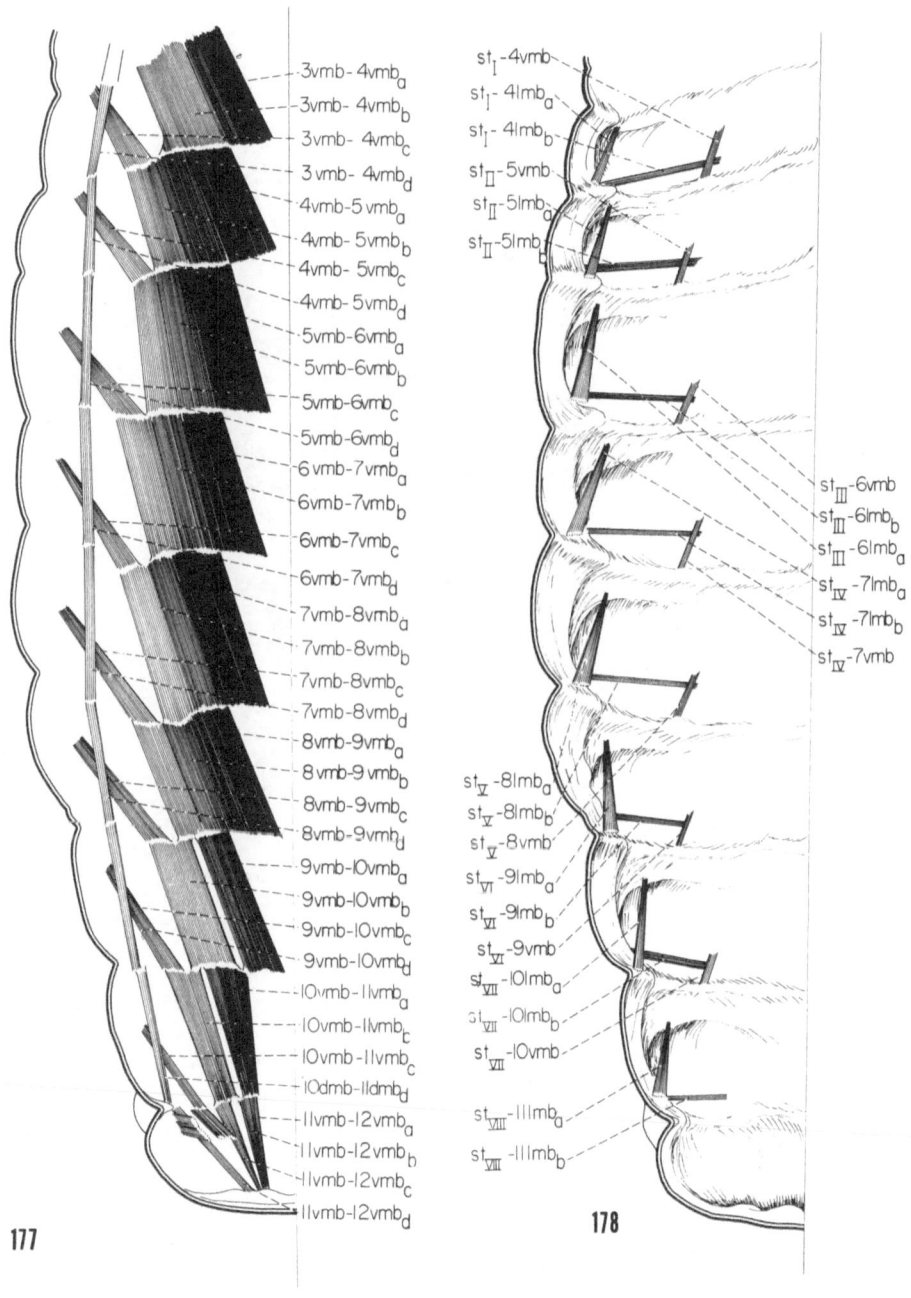

177

- 3vmb - 4vmb$_a$
- 3vmb - 4vmb$_b$
- 3vmb - 4vmb$_c$
- 3vmb - 4vmb$_d$
- 4vmb - 5vmb$_a$
- 4vmb - 5vmb$_b$
- 4vmb - 5vmb$_c$
- 4vmb - 5vmb$_d$
- 5vmb - 6vmb$_a$
- 5vmb - 6vmb$_b$
- 5vmb - 6vmb$_c$
- 5vmb - 6vmb$_d$
- 6vmb - 7vmb$_a$
- 6vmb - 7vmb$_b$
- 6vmb - 7vmb$_c$
- 6vmb - 7vmb$_d$
- 7vmb - 8vmb$_a$
- 7vmb - 8vmb$_b$
- 7vmb - 8vmb$_c$
- 7vmb - 8vmb$_d$
- 8vmb - 9vmb$_a$
- 8vmb - 9vmb$_b$
- 8vmb - 9vmb$_c$
- 8vmb - 9vmb$_d$
- 9vmb - 10vmb$_a$
- 9vmb - 10vmb$_b$
- 9vmb - 10vmb$_c$
- 9vmb - 10vmb$_d$
- 10vmb - 11vmb$_a$
- 10vmb - 11vmb$_c$
- 10vmb - 11vmb$_c$
- 10dmb - 11dmb$_d$
- 11vmb - 12vmb$_a$
- 11vmb - 12vmb$_b$
- 11vmb - 12vmb$_c$
- 11vmb - 12vmb$_d$

178

- st$_I$ - 4vmb
- st$_I$ - 4lmb$_a$
- st$_I$ - 4lmb$_b$
- st$_{II}$ - 5vmb
- st$_{II}$ - 5lmb$_a$
- st$_{II}$ - 5lmb$_b$
- st$_{III}$ - 6vmb
- st$_{III}$ - 6lmb$_b$
- st$_{III}$ - 6lmb$_a$
- st$_{IV}$ - 7lmb$_a$
- st$_{IV}$ - 7lmb$_b$
- st$_{IV}$ - 7vmb
- st$_V$ - 8lmb$_a$
- st$_V$ - 8lmb$_b$
- st$_V$ - 8vmb
- st$_{VI}$ - 9lmb$_a$
- st$_{VI}$ - 9lmb$_b$
- st$_{VI}$ - 9vmb
- st$_{VII}$ - 10lmb$_a$
- st$_{VII}$ - 10lmb$_b$
- st$_{VII}$ - 10vmb
- st$_{VIII}$ - 11lmb$_a$
- st$_{VIII}$ - 11lmb$_b$

Figs. 177–178. Abdomen of SG larva. Dorsal view (left half) (series concluded).

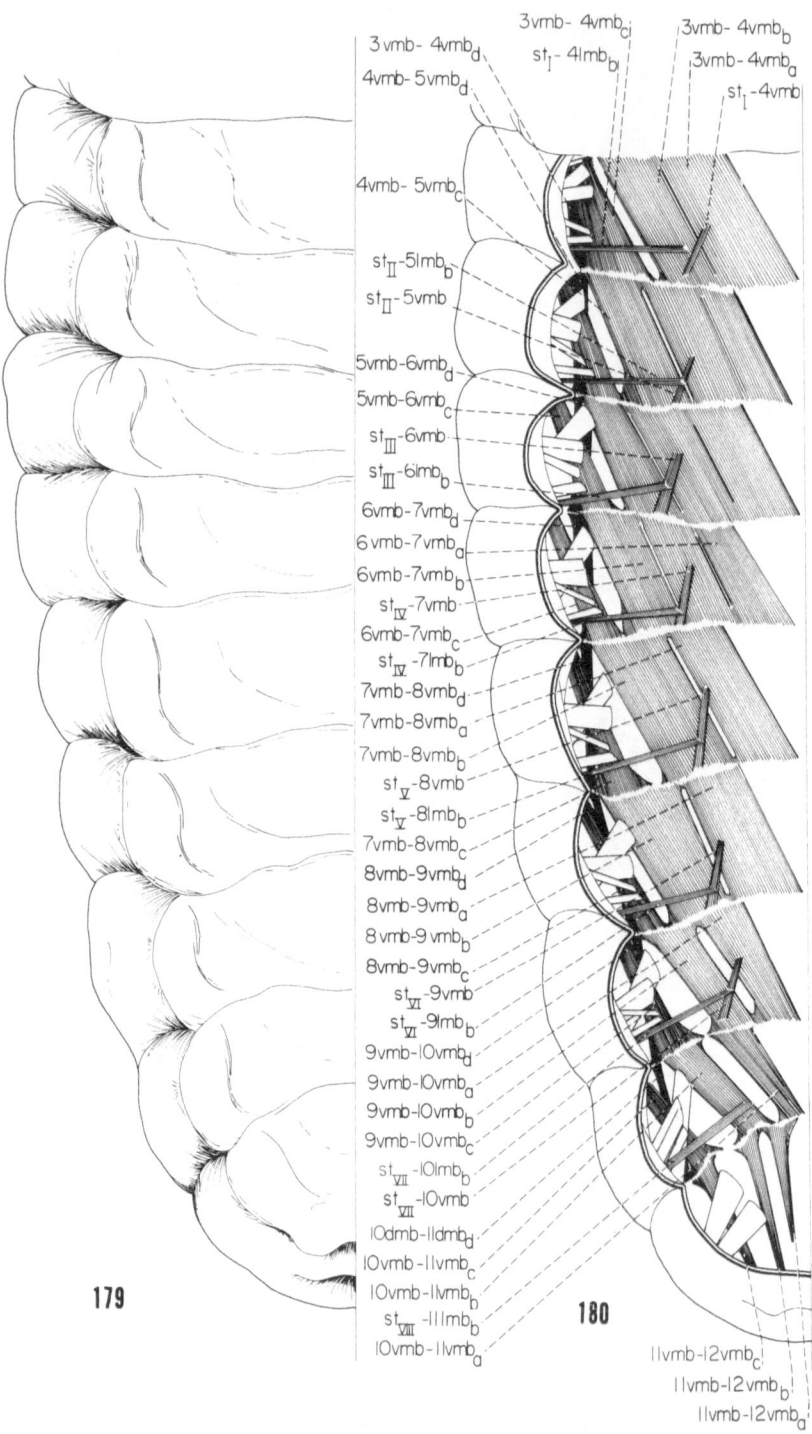

179

3vmb- 4vmb$_d$

4vmb- 5vmb$_d$

3vmb- 4vmb$_c$

st$_I$- 4lmb$_b$

3vmb- 4vmb$_b$

3vmb- 4vmb$_a$

st$_I$ -4vmb

4vmb- 5vmb$_c$

st$_{II}$-5lmb$_b$

st$_{II}$- 5vmb

5vmb-6vmb$_d$

5vmb-6vmb$_c$

st$_{III}$-6vmb

st$_{III}$-6lmb$_b$

6vmb-7vmb$_d$

6 vmb-7vmb$_a$

6vmb-7vmb$_b$

st$_{IV}$-7vmb

6vmb-7vmb$_c$

st$_{IV}$ -7lmb$_b$

7vmb-8vmb$_d$

7vmb-8vmb$_a$

7vmb-8vmb$_b$

st$_V$-8vmb

st$_V$-8lmb$_b$

7vmb-8vmb$_c$

8vmb-9vmb$_d$

8vmb-9vmb$_a$

8 vmb-9 vmb$_b$

8vmb-9vmb$_c$

st$_{VI}$-9vmb

st$_{VI}$-9lmb$_b$

9vmb-10vmb$_d$

9vmb-10vmb$_a$

9vmb-10vmb$_b$

9vmb-10vmb$_c$

st$_{VII}$-10lmb$_b$

st$_{VII}$-10vmb

10dmb-11dmb$_d$

10vmb-11vmb$_c$

10vmb-11vmb$_b$

st$_{VIII}$-11lmb$_b$

10vmb-11vmb$_a$

11vmb-12vmb$_c$

11vmb-12vmb$_b$

11vmb-12vmb$_a$

180

Figs. 179–180. Abdomen of SG larva. Ventral view (left half) (see also Figs. 181–184).

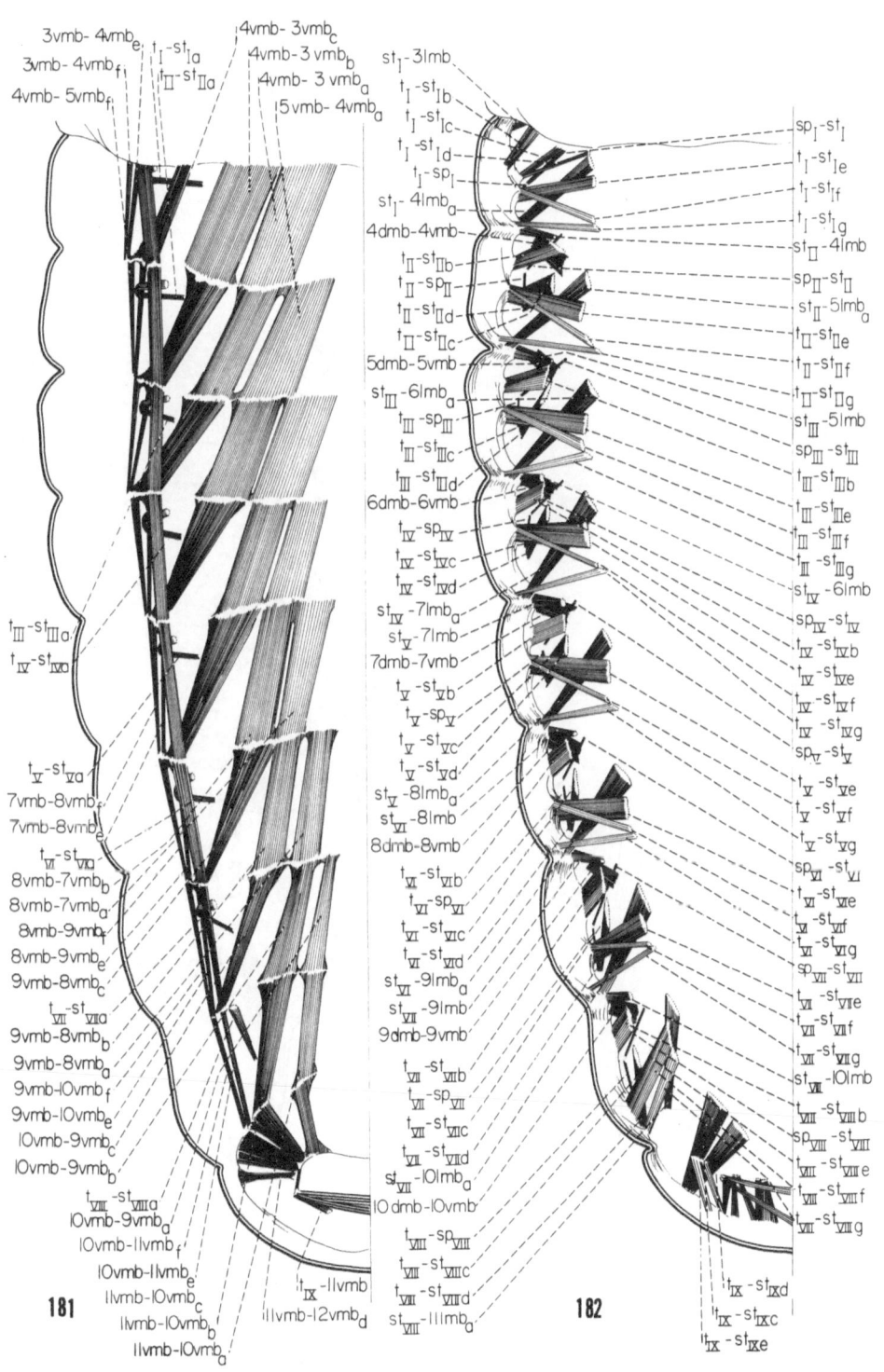

Figs. 181–182 Abdomen of SG larva. Ventral view (right half) (series continued).

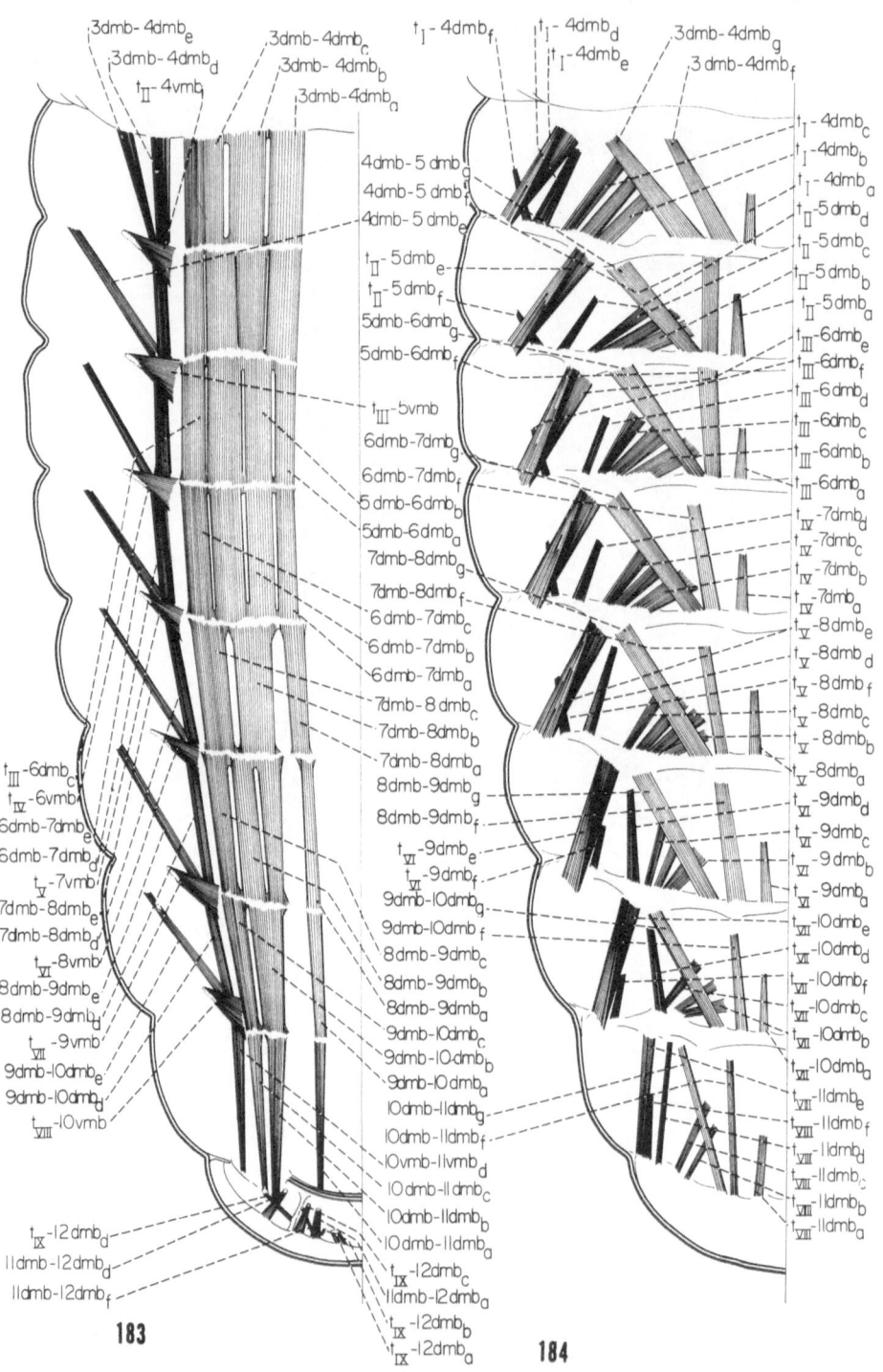

Figs. 183–184. Abdomen of SG larva. Ventral view (right half) (series concluded).